Przemysław Busse, Włodzimierz Meissner

Bird Ringing Station Manual

Hand drawings: Tomasz Cofta

Managing Editor: Katarzyna Michalczyk

Language Editor: Kristina Kangas

DE GRUYTER OPEN

Published by De Gruyter Open Ltd, Warsaw/Berlin
Part of Walter de Gruyter GmbH, Berlin/Munich/Boston

ISBN: 978-83-7656-052-6
e-ISBN: 978-83-7656-053-3

Managing Editor: Katarzyna Michalczyk
Language Editor: Kristina Kangas

www.degruyteropen.com

Cover illustration: © Katarzyna Stępniewska

Przemysław Busse, Włodzimierz Meissner

Bird Ringing Station Manual

Motto: "In God we trust... everyone else must bring data" - from Evan Cooch's mail

Contents

Preface

Everybody interested in birds – an ornithologist, a recreational birder, and even bird afficionados know about bird ringing, or banding. For most, ringing is only a fragment of their general education. However, a subset of people who learned about ringing were so excited with the method, and how it could be used to solve the bird migration problem, that it became the core of their professional life. This was the case of the authors of this manual. The older one of us – Przemysław Busse – ringed his first five chicks in a swallow nest as a 16 year-old boy in 1953. The subsequent year, he was accepted as a ringer, and in 1955, he started to ring chicks of Black-headed Gull, White Stork and swallows as a member of the Polish Ringing Centre expeditions to Masurian Lakes district. Ringing of full-grown birds was a great discovery during a visit, with Bob Spencer as a host, at the Bradwell Bird Observatory in Great Britain in 1959. The result was surprising. In 1960, the first Polish bird station tried pioneering ringing of full-grown birds during autumn migration, and a group of students from the University of Warsaw started to learn bird migration at the Polish Baltic coast. Our minds were fresh, for we knew nothing about bird migration, but we were enthusiastic. Since bird migration is a vast and very complicated phenomenon, the main idea became to create "*a NETWORK working according to the standard methods*". We dreamt about long-term study programme, which later manifested into "Operation Baltic", which is still the longest-running bird migration research network. The saying *"Never laugh at anyone's dreams. People who don't have dreams don't have much"* was confirmed, and we have the long-term programme to prove it. Among the Polish ornithologists gaining interest in birds during the 1960s and 70s, there was not a single person who did not pass through the Operation Baltic experience. The second of listed authors is not the exception. Although his time with the Operation Baltic work was short (only few days), he worked at the station in southern Poland that organized the ringing of waders. This was enough to establish the ringing station at mouth of the Reda river devoted to wader study in 1983, and it followed suit in the maintenance of long-term studies. The group was called KULING, and it is still actively attracting new generations of ornithologists.

During many years of intensive work at Operation Baltic and KULING, many methods and ideas from the field study of migrating birds were tested. We gained a great deal of experience. Political borders were broken, and the saying that "*birds do not know boundaries*" could be followed by students of bird migration – that is truly a phenomenon on the continental scale. While visiting many bird stations, from Spain and Britain to Russia and from Finland and Sweden to the Middle East and even Tanzania and South Africa, we found that the main problem of international co-operation is the incompatibility of methods and the local routines of work, and these make exchange and efficient use of data files very difficult. And once more the idea of "*a NETWORK working according to the standard methods*" appeared to be a proper solution. In 1996, an international network was arranged on the East

European flyway – the SEEN (SE European Bird Migration Network). This gave us the opportunity to collect broader experiences and confront new exciting problems to study. The manual is intended to collect and present to wider audience more than 50 years of practice in the field work of ringing of birds, mainly during migration.

We are deeply grateful to all, for there are numerous people who contributed to this manual and discussed its contents. Especially, we would like to thank to Dr. Vladimir Payevsky, Russia and Dr. Ricardas Patapavicius, Lithuania for supplying us with details about working with big Heligoland traps that were less known to us.

We hope that the manual will be useful, not only for bird stations, but for many ringers who ring birds not only for fun, but for whom the ambition is to contribute to science as well. Obviously, not all parts of the manual will be applicable in an individual ringing, but it is always worth it to contemplate anew upon the encounter of novel, sometimes apparently strange ideas – especially when we plan to start a new project.

Przemysław Busse, Włodzimierz Meissner

Figure P1: Three generations of Operation Baltic – KULING – SEEN ringers: Przemysław Busse, Włodzimierz Meissner and Magdalena Remisiewicz.

50 Years of OPERATION BALTIC

Figure P2: Logo of Operation Baltic (by J. Desselberger and P. Busse).

The Operation Baltic is a scientific programme with the aim of long-term, complex bird migration study along with monitoring numbers of migrating birds. The programme has been running continuously since 1961 and covers both autumn and spring migrations.

The fieldwork of the programme is carried out at few (2-6) seasonal bird stations situated along the Polish Baltic coast.

The routine work consists of:

1. Mist-netting of birds with stable number of nets, working continuously during migration periods (March 24th - May 15th and August 14th - November 2nd). Standardised catching allows evaluation of long-term number trends in the bird populations passing the stations;

2. Ringing and measuring of all individuals caught. The standard set of measurements contains: wing-length, tail-length, quantitative wing-formula, weight, and fatness. This is the basis for migration and biometric studies;

3. Visual observations of diurnal passage; bird species, numbers and direction of flight are observed all day, fifteen minutes per hour; this gives an information about the dynamics of migration as well as long-term trends in species not caught by the nets;

4. Extensive testing of nocturnal migrants for directional preferences using new orientation cage desing which is one of the most important points of the fieldwork now;

5. Volunteers. University students and amateur birdwatchers, who help the professional staff of the stations, do most of the fieldwork. The collaboration with amateurs allows around 50 young people to be trained in bird identification every year, along with increasing exposure to scientific work and raising awareness about the bird protection. Participation of volunteers makes the cost of programme work relatively affordable.

During Operation Baltic, over 1 600 000 birds were ringed and over 1 000 000 of them fully measured. For few small passerines, e.g. Goldcrests, collected data exceeded European totals. A dozen or so thousand ringing recoveries cover Europe, from Portugal to the Ural Mountains and from Finland to Greece, as well as few in Middle East and African countries. It can help to look for areas that are critical for bird survival. This is very important when monitoring data are evaluated.

The Operation Baltic monitoring data represent one of the longest passerine monitoring series in the world. They are also the most extensive, as they contain information from few stations working according to the same methodical standard, and they cover a much wider variety of species (parallel bird netting and visual observations) compared to stations limited to bird catching only. No single passerine bird species is excluded from the field of interest of the programme, so the Operation Baltic data contain information about a number of rare, even endangered, species of birds.

The value of monitoring data grows exponentially with prolongation of the data series. Evaluation of long-term population cycles is possible only when periods of monitoring work exceed the length of the cycle. The level of short-term fluctuations around the long-term trends can be studied successfully only on long data series.

30 Years of WATERBIRD RESEARCH GROUP KULING

Figure P3: Logo of Waterbird Research Group KULING (by M. Skakuj).

The story of KULING has its roots since July 1981. During this season, two students from University of Gdańsk had seen walk-in traps for the first time in their lives and took part in wader ringing. After a couple of days spent at a wader ringing camp in southern Poland, they returned to the Baltic coast with the idea of organizing wader ringing camp in the Reda river mouth. The Waterbird Research Group KULING was established in 1983 by a group of so-called KULING fathers (Włodzimierz Meissner, Bogdan Brewka, Michał Skakuj and Arkadiusz Sikora), as an informal part of the students' scientific circle dealing with autumn migration of waterbirds and waders in the Gulf of Gdańsk.

Since the very beginning, KULING had a marvellous atmosphere that led students, pupils, as well as amateurs and professional ornithologists to work together in the field and share ideas during social meetings. KULING's field activities became focused on two main topics: studies on wader migration and monitoring number of waterbirds during the non-breeding season in the western part of the Gulf of Gdańsk. In 1996, WRG KULING was registered as a non-governmental organization, even though it is closely connected with University of Gdańsk.

Between 1983 and 2012 KULING team ringed more than 70 000 waders, about 8 000 wagtails and pipits, and more than 8 000 gulls. The vast majority of birds were ringed at temporary working ringing stations. The overall number of birds ringed by KULING reached 100 000, and this gives to us the opinion that it is one of the most effective ringing teams in Poland.

The archive of publications signed as "paper of WRG KULING" consists of 150 scientific papers, and the majority of them concerns wader migration. Notably, we are also involved in educational activities in the Polish coast.

We decided to call our group KULING after the name used by the native people of the Puck Bay area for large waders. They recognize only two kinds of waders: small ones (up to Knot, *Calidris canutus*, size) that they call "bigus", and large ones that they call "kuling". It sounds very similar to "kulik" – the Polish name of the Eurasian Curlew, *Numenius arquata*. "Kuling" seemed to us a better name than "bigus", and we adopted it as the name of the group.

15 Years of SE EUROPEAN BIRD MIGRATION NETWORK

Figure P4: Logo of SEEN (by T. Cofta and P. Busse).

Bird migration is a phenomenon on a continental scale, greatly differentiated in various regions, and it has many complicated features. There are some main flyways within continental migration systems that are studied more or less in detail. Bird populations of the same species, but originated from different areas, can migrate *via* various flyways and to different winter quarters. So, it is not enough to study the phenomenon at a single bird station, at one limited area, nor along a single flyway. The "network style of work" is absolutely necessary. Some attempts to network at an international scale were run by Operation Baltic, which attracted a few ornithologists from the late Soviet empire to collaborate. It was a very limited collaboration. Next, trials were made within "ESF European-African Songbird Migration Network", working for three years on the SW bird migration flyway in the mid-1990s, and finally *Manual of Field Methods* was published. In 1996, a group of ornithologists from northern and central Europe established SEEN ("SE European Bird Migration Network") that focused on the SE flyway that had been poorly studied as of yet. Necessity for a common methodical program led to the preparation of a comprehensive manual, published in 2000, which gave not only methods of the fieldwork, but also methods of evaluating the collected material.

SEEN is an international umbrella organization of institutions studying bird migration along the South Eastern migration route that leads from Europe and western Asia to Africa. Our objectives are to encourage research and enhance understanding of migratory flyways, to establish a uniform methodology of bird migration data collection, to elaborate and develop new techniques of data analysis (including computer software, highly specialized statistical methods etc.), to assist in international co-operation and the exchange of information and experience relating to bird migration, and finally, to promote the conservation of birds and their habitats. Our network offers support in organizing bird migration research, regularly conducting training in the methodology of bird migration studies, particularly in countries where, until now, such studies were either not conducted at all or conducted on a very irregular basis. Participation in SEEN is non-exclusive. Ornithological research stations, departments and laboratories of colleges, universities and other schools of higher learning, scientific and nature protection organizations, non-governmental organizations and study groups working in the field of bird migration research can become members of SEEN.

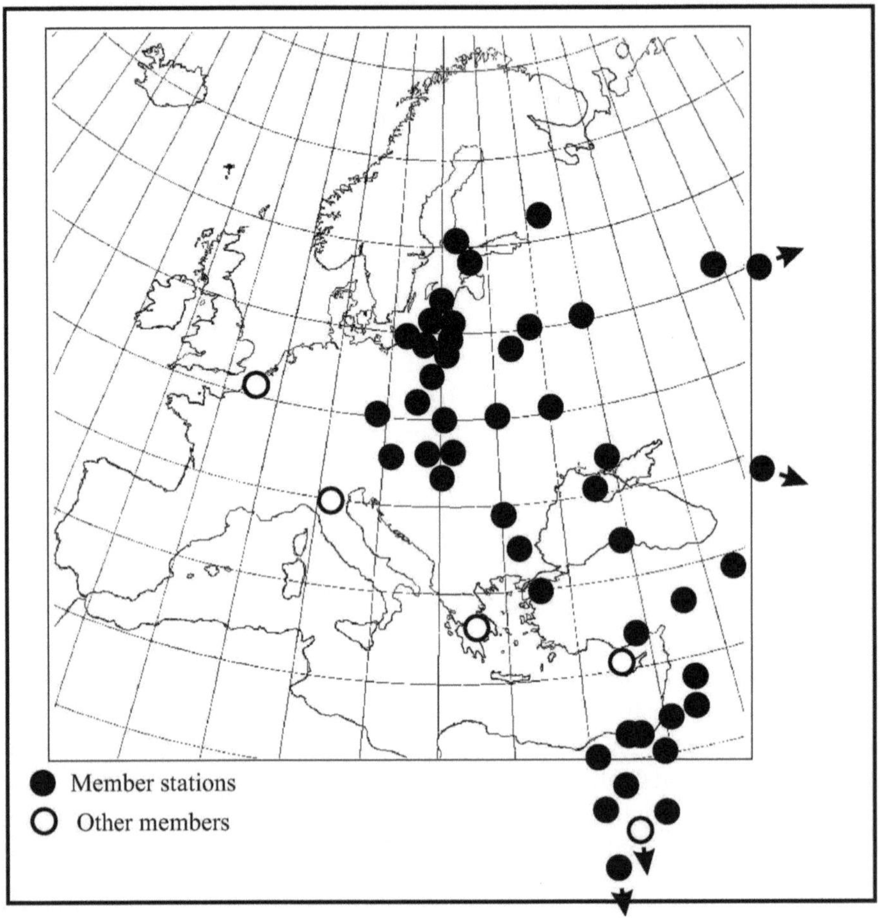

Figure P5: The map of the SEEN sites.

Introduction

Collaboration between bird ringing stations of a research network requires both standardization and flexibility. The aim must be to standardize elements of the station routine, where results will be directly compared during further evaluation of data. This includes: techniques for measurements, orientation experiments, or monitoring. On the other hand, flexibility should allow collection of standard data as well as different specific studies performed within our own projects and local agreements with various partners. One of the most important tasks to setting up the station routine is organizing the work in such a manner that an optimum output of results will be obtained with a minimum effort. Optimum results mean not only a maximum number of birds caught, but also the collection of useful, scientific data with the sources at hand. Depending on local conditions, catching devices, and the size of station staff, and optimisation of a station routine may need more or less attention. At any rate, it will make the work easier, more effective and satisfying. So, let us try to establish the station routine in a way that is favourable to both birds and ringers! All people catching and ringing birds use some of these working methods, while others turn out to be more habitat- or bird-group specific. In this book, methods will be presented for catching passerines with mist-nets on land and in wetland habitats. Furthermore, this manual will explain how to work with Heligoland traps, and, finally, how to catch waders and, to a limited extent, raptors/owls with nets and traps. At times, the methods described will be applicable to catching birds from other groups, but these possibilities are taken into consideration here. According to the main focus of a ringing station, two main types may be discerned: "passerine" and "wader" stations. Most of the chapters will contain information common to both types.

PART I: **The Passerine Station**

1 Methods of the Field Work

1.1 Catching

The point of departure for the standard of all bird ringing station work is the agreed-upon number and the quality of catching devices. In most cases, mist-nets are used, and their number and construction determine how to stabilize the catching effort both on a seasonal as well as a long-term scale. If a Heligoland trap is in use, the only standardization problem is the operation time of the trap, which will be influenced by wind force. In modern migration research, the dynamics of seasonal bird migration is the basis for the interpretation of other data. Therefore, stable operating time of Heligoland traps or a standardized number of nets used during the season are an essential methodical requirement. The number of nets in use must be fixed to a level at which the number of available staff can safely handle all birds caught. The main aims of bird ringing station work are: monitoring of bird numbers and seasonal migration dynamics and collecting data for biometrical and other studies of special objectives. Rarely, the work at the ringing station is dedicated only to ring maximum possible numbers of birds.

According to the aims of work at a station, the catching methods must fulfil some requirements (Table 1.1).

Table 1.1: Constrains for different kinds of studies.

Aim of the study	Number of nets
1. Monitoring	stable within a season, stable between years
2. Seasonal dynamics	stable within a season
3. Biometrics	recommended stable within a season
4. Special studies	recommended stable within a season
5. Ringing only	allowed variable number*

* but see p. 4 (point 5)

These requirements can be listed more in detail:

1. Collecting of monitoring data

In this case, the highest level of standardisation is necessary.

1.1. The work must be planned for a sequence of years.

1.2. Time and period of work is standardized. Within the season, work should be carried out continuously, or at least be made as regular and frequent sampling (this compromise, however, is not recommended!). This is because the migration intensity is very irregular – after a day with no birds, one can have a rush of hundreds or thousands and more birds caught. In the Operation Baltic practice, we had even days when around 20% of yearly catches of one species occurred. Missing such days when sampling could drastically change the total value for the season.

1.3. Equal numbers and quality of nets should be used from year to year, and when new nets are added to standard set, birds caught in the added nets should be treated separately. It must be stressed, however, that any changes affect comparability.

1.3.1. The number of nets should be stable within a season (minor changes may be compensated for when data are evaluated). The nets damaged or stolen should be replaced as soon as possible; good hint: have a stock of a few nets at hand as a backup for replacement.

1.3.2. The daily netting routine should be stable. It is advised to catch birds continuously without closing the nets for night time – in many places, catching peaks do not occur regularly at the same time of the day; e.g. thrushes that have landed after a sea-crossing start to be active in the middle of the day, instead of early in the morning as usual. If possible, do not close nets during migration peaks (unless survival of birds caught is endangered – but also see *Laboratory Working Routine* hints – p. 118). At some sites, because of special constraints (e.g. high temperatures and insolation in lower latitudes, known and very stable daily catching pattern, and living conditions of the staff), nets may be closed for part of the day. It is advisable to do this regularly at the same time by not prolonging catching because e.g. "there are a lot of birds today", or in order to finish work earlier as it seems a "poor day"). Within Middle East and Northern Africa, a good time for a siesta is between 11.00 and 15.00-16.00.

1.4. Changes of environment must be taken into consideration. Three ways of minimizing the influence from such changes can be listed – (1) arranging the catching area within a relatively stable environment (such habitats are, however, usually not very rich in bird species), (2) controlling growth of trees and bushes (note, however, that the surrounding area will be changing all the time), (3) actively shifting the catching plot within a bigger area of similar environmental conditions (value to birds!). A combination of these methods could be applied according to knowledge of local conditions.

1.5. The nets should be located in different habitats and the distribution of nets relative to habitats ought to be stable over the years.

2. Seasonal dynamics of migration

This is one of the most important types of data in any context.

The contents of points 1.2 - 1.3 (above) should be attended to, but any sampling error may affect the picture of seasonal dynamics very much. It is important to remember that during one missing day up to 20 percent of the annual catch of one species could be missed!

3. Bird measurements

Catch as many birds as you are able to measure, but note: bird measurements without the possibility to localise these measured birds within their migration waves have very limited value! Therefore, adjust number of permanently opened nets to the expected high level of catching (but not to single peaks).

4. Orientation tests, blood sampling, parasite sampling etc.
Catch as many birds as you are able to handle with these techniques, but recall previous note under point 3.

5. Ringing only
Catch as many birds as you are able to ring (including sex/age determination!). Erect as many nets as you are able to handle without bird losses; eventually use tape-luring, however remember – and once more remember – that station work is not a ringing championship, but means the collecting of scientific data. Today, ringing of migrants is closely connected with collection of other types of data. Seasonal dynamics must be known when ringing recoveries are evaluated in the modern way. Reduce your order for nets, unless they are planned for storage a reserve for replacements, and try to fulfil requirements stated under point 2.

1.2 Visual Observations

Visual observations are frequently performed at ringing stations. They are focused on different groups of species according to the main field of interest of the station staff. Bird counts may be performed in two different ways – (1) counting birds in active migration flights and (2) counting those resting in surroundings of the station area. The first method is used mainly at the "passerine" stations, where both passerines and other diurnal migrants are counted there, as well. The second method is used at "wader" stations, however it is not possible to follow all wader migration as they migrate mostly by night – for more details see *Wader Counts* - p. 160). In many localities situated at guiding lines like sea coast, spits, and rivers, the stream of diurnal migrants follows a well-defined course, and this may fluctuate within very narrow limits. At other sites, diurnal migration will show "broad-front" character and migrating birds will be dispersed over the whole area. In the first case, the migration count will be more effective, as birds are observed even if actual migration is not intensive. On the other hand, the count could be difficult during a mass passage when there are tens of thousands of migrants per day. Out of concentrated streams of migration, visual observations could be boring, as a low number of migrants are observed, but even in such cases, one could collect interesting data. Areas with no clear guidelines may give biased results due to local changes in the concentration of flying birds, e.g. when parts of flocks may stay outside observers' sight.

Visual observation of the passage should be made from a fixed stand located at the local stream of migration, if there is one within the station area. In order to get good estimation of the total number of birds passing the observation point, all day observations should be applied, especially in localities where the intensity of the passage notably varies during the course of a day – this frequently occurs at the sea coast, where some birds have crossed the sea prior to reaching the local stream of migration. Usually, when birds migrate over land, the passage is limited to a few hours after sunrise. In some coastal areas, peaks of diurnal passage occur around noon or even in the afternoon. It is true, however, that observations made during out of the peak of passage are tiresome and boring to the observer, unless they are given a chance to rest. Because of this the observation time during lack of passage can be shortened.

There are two methodical variants of the migration counts used:
1. Continuous observations from sunrise to sunset, or at least for 6-8 hours. Observations are performed on a **daily scale** throughout the migration period, which is a difficult task, but the result will be the real number of birds passing by an observation point,
2. Sampling observations done on a **daily scale** and within a day where a sampling procedure is applied (usually 15 minutes per hour) – this method allows estimates of real numbers of birds passing and is not equally exhausting to the observer. Moreover, it does not disturb ringing at the station, because counts are conducted at the time of mist-net checking. The correlation of the results with the first method is at the level 0.90 (after own comparisons of these two methods in early years of the Operation Baltic studies), which can be accepted as very good.

It is critical to stress that, as in catching, any sampling not based on daily counts is biased by the continuously changing migration dynamics of every species, including both diurnal and nocturnal migrants.

The recommended observation routine (Fig. 1.1):
1. The observations of the passage are carried out at 15 min per hour sampling, starting at full hours, and beginning around sunrise and continued till sunset. When there is no observable migration in two consecutive 15 min observations, the next observation is shortened to 5 min; return to the normal routine must be applied consistently when the observer notices intensification of the passage of at least one species. If it is evident after a few years of observation, that the particular locality has no noon and afternoon movements, visual observations might be limited to highly effective times only.
2. At places where intensive bird migration occurs, birds are identified, by sight and sound, and counted within a flexible range – for small birds the range should allow for the identification and count by means of the naked eye, without use of binoculars (when many birds pass, there is no time to check all birds with binoculars); in larger birds (e.g. when raptors are included), the range is limited to a sector within which it is possible to see the bird with the naked eye, but

where identification is made with the use of binoculars. It is advisable to fix the observation point at such a place, that most of the birds migrating within the local stream pass the observer to the north and west (as they are visible in a better light). As the local stream may shift a little with wind direction and force, it is advisable to shift the observation stand within 100 m relative to the standard point, adjusting the actual place to better visibility, since the birds passing between the observer and the sun will hardly be identifiable.

3. The birds are noted in a note-book, listing their species name by code – see *Species Determination and Coding* – p. 81 - in observations (5–letter code is more convenient than the 6-letter one), direction of flight (by wind-rose, 8 directions) and number; birds flying in the most commonly observed direction of the passage (standard direction must be specified at the beginning of the note-book) can be noted as numbers without the direction letters (e.g. CASPI 40, 10, 50...) – all others must be accompanied by the letters describing direction (e.g. SE 30, N 25...), but when non-standard direction is repeated, a bracket could be used (e.g. SE [30, 20, 5], E 15...). For standard visual observations, when only total numbers of individuals per species is needed, the subsequent numbers do not describe the size of the flocks passing the observation post, but may be accumulated values for a couple of flocks pooled together. For instance, CASPI 50 does not necessarily mean "a flock of 50 siskins", but could mean: four Siskin flocks: 10, 30, 5, 5, adding up to 50. In this way, notation will be quicker, which is important when a lot of birds migrate. When flock size is wanted, it should be clearly stated in the local instruction.

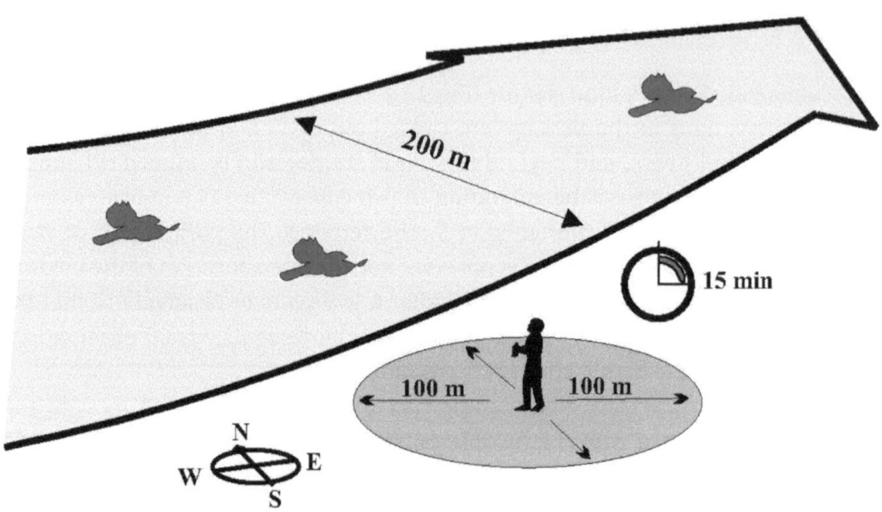

Figure 1.1: Visual observations' routine (see text for explanation).

A basic rule worth noting is that the same individual should make both observations and notations – there is standardised missing of the birds which pass the point. Noting by dictation to a digital recorder or noting by another person changes detectability of birds and, if applied, it must be used for all observations performed, because of compatibility reasons.

Visual observation of the resting birds is usually performed at the "wader stations" (see *Wader Counts* - p. 160).

2 Passerine Station Field Equipment

2.1 Mist-nets

There are different types of mist-nets in use, but generally they are as shown at Figure 2.1.

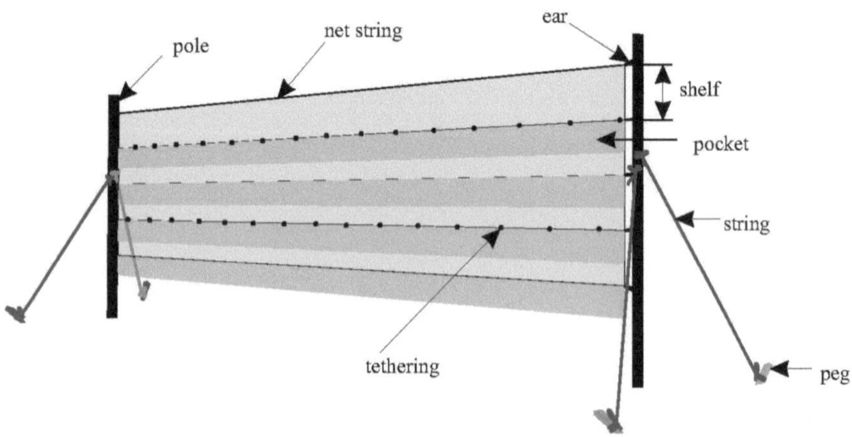

Figure 2.1: Mist-net. Basic terms.

The main parameters describing them are:

1. **Thread used.** Contemporary nets are made from nylon, polyester or some other similar synthetic thread. The material and finishing treatment determine softness or hardness of netting and its UV resistance. UV rays destroy the netting material by causing total damage to the net. The most UV sensitive are nylon nets. Some of the synthetic nets are very hard and may cut the bird's skin when it becomes heavily entangled. Nylon netting is much softer than polyester one. The potential danger of the net to birds should be treated as one of the most important characters when net types are chosen. This feature is strictly connected with the thickness of the thread, which is characterized by the "denier" measure (weight in grams of 9000 meter thread) and the "ply" (the number of threads twined), e.g. 50d/2, 70d/2, 110d/2, 235d/2 etc. Thinner thread means lower visibility, better catching ability, higher degree of entangling of birds (they are difficult to remove – the time spent on removing will be longer), and ultimately much higher probability of skin and feather damage to birds. In addition, there will be more holes made by twigs, thorns or heavy birds caught, as

well as lower UV resistance and greater laboriousness of net cleaning. Thicker thread, in turn, means lower catching ability of the net. However, birds are not entangled and are easier to remove, thus, saving time. With such nets, beginners are less likely to injure the bird. Cleaning the thicker net from leaves is much simpler and the procedure is safe for the net. Nets of this kind also have high durability because there will be fewer holes caused by entangling of bushes and catching heavy birds. In addition, the netting has much higher UV-resistance.

Thin nets are recommended only when the catching area is very open and the aim is to be very efficient in catching particular bird species living in such an environment (e.g. swallows, stonechats, wagtails etc.). Furthermore, this particularly applies when there are not too many birds to catch, when the staff consists of well trained ringers, and the station routine includes only few studies. On the other hand, use of thin nets is not advisable when catching is done in areas where mass migration could be expected. Thin nets are much more expensive as well, especially when one takes under consideration high turnover rate of such nets.

Thick nets are recommended when the catching area includes more dense vegetation, high number of birds during peak days, untrained helpers to remove birds, and if the station routine includes a detailed examination of the birds. Effective netting with thicker nets necessitates the use of more nets, which in turn means bigger effort to erect them, but much lower turnover rate of the nets makes the mist netting cheaper. Special nets are produced and used, which are made from elastic thread that in normal state, i.e. without birds caught, have no shelves. Individual "pockets" are made by the birds caught. Conversely, monofilament threads used for production of some nets are less soft and less visible than normal thread nets, but usually they are more dangerous to the birds caught.

2. **Mesh size.** This parameter is given in two different ways: "knot to knot" and "stretched". E.g. 16 mm knot-to-knot size is equal to 32 mm stretched (Figure 2.2). Here we will use "knot to knot" measures. The mesh sizes used are differentiated depending on the species for which the effective use of the net is intended. For small passerines, the mesh size most in use is 16 or 17 mm. It is small enough even for mass catching of Goldcrests. Smaller sizes (14-15 mm) have lower catching ability. Sixteen millimetres mesh net has lower catching ability when bigger birds are involved (the size of thrushes or larger). In contrast, many small birds (as Goldcrest, leaf warblers, Reed Warbler etc.) could easily pass through 18 mm mesh. Small birds will usually get much more entangled when caught in such a net. A special problem with 18 mm mesh size arises when large amounts of starlings get caught; the 18 mm mesh just fits to the bend of the Starling wing and its first primary works as a fish-hook, thus, removing the wing frequently causes injury to the bird. Thrushes accidentally caught into raptor nets (mesh sizes 45-80 mm) may get very entangled. Thus, we recommended 16 mm mesh size as a standard in passerine station, despite some small or slim bodied birds can pass through of them.

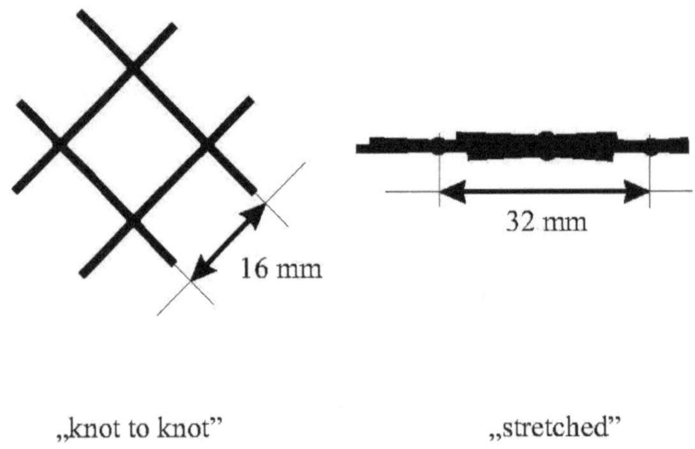

Figure 2.2: Methods of description of the mesh size.

3. **Number of shelves.** As a standard, nets usually have four-five shelves. Previously, the standard was three shelves and some traditionalists still use this type. Some special nets with less (1-2; "pipit nets") or more shelves are also in use. The number of shelves should be considered in connection with the height of each shelf. The most efficient shelf height is approximately 50 cm. Broader shelves, paradoxically, do not have higher catching ability, since the upper part of the shelf works as a dead area, from which the bird is deflected. Only a fraction of such birds will return once more to the net and get caught in another shelf. So, the optimal number of shelves may be fixed as four or five, giving maximum catching ability as well as easy and quick removal of birds from the net. Many-shelf nets were specially developed and need to be pulled down when a bird is caught in the uppermost shelves, so they are not recommended as standard equipment at the station, especially when higher numbers of birds are expected to be caught. Such nets also make monitoring comparisons complicated (birds from them should be noted separately). In addition, one- or two-shelf special nets, sometimes used for catching birds at meadows and marshes are not recommended as a supplement to the standard set of nets used for monitoring purposes. Four one-shelf nets are not equal in efficiency to one four-shelf net of the same length.

4. **Net dimensions.** The height of the net depends on the number of shelves and their height – this was discussed above. The length of nets in use may vary. Most commonly used are nets of 6-7-10-12-14-18 m lengths, at times even longer. Shorter nets fit better to special locations, like the front of small bushes, across ditches, etc. They also may be used in order to create long rows of nets of complicated curvature adjusted to paths of vegetation. Long, straight rows made of longer nets are more economical, considering

the number of poles needed to erect them. However, longer nets are more sensitive to wet and windy weather. Wet nets become longer and heavier and frequently even touch the ground. They will easily pick up surrounding vegetation when there is a wind. Some nets are produced using water resistant strings; they hold the same length when dry and wet. This is a great advantage, especially when long lines are used. However, nets that are still wet will have deeper pockets that can touch the ground below the net or entangle twigs of surrounding bushes. The catching value of one, say, 14 meter-long net is not exactly equal to two 7 meter-long nets, but their summation may be considered approximately equivalent. It is advisable to use only two lengths of nets: short net (6-7 or 10 m) and a doubled-length net (12-14 or 20 m, respectively). Recalculations of the catching results, e.g. per 100 m of nets, when more types of nets are used, are less precise. This is because the catching efficiency across the net is not the same: the middle has a higher catching efficiency than the ends attached to the poles. Since it is rather lower, while in two nets have four "ends", this is double the number compared to a single longer net with only two "ends," and this will impact the overall catching efficiency, despite having approximately equivalent lengths.

5. **Tethering (wind blockades).** Tethering means that netting is fixed to the horizontal net strings with an additional thread in order to prevent the netting from slipping along the string when there is a wind blowing parallel to the net. Good tethering is important when many birds of average size (e.g. tits, chaffinches compare Figure 17.2-1) are caught simultaneously, for it prevents birds from clumping together in one corner and further impeding the catching efficiency of the rest of the net. One-line tethering is definitely not enough to effectively protect the net from wind. For a short 4- and 5-shelf net, optimal tethering should be doubled at the second and fourth string. A tethering of this kind is symmetrical, so there is no need to take a blocked string on top of the net when erected (as is the case with only one tethering). Triple tethering is advisable for long nets and special ones for raptors or thrushes.

6. **Net colour.** As a standard, black nets are used. However, the colour of the net is not as important as many ringers think when looking for nets that do not change colour with extended use (not bleaching), especially in very rainy or, inversely, very sunny regions. In practice, according to our experience, green or brownish, even sandy colour nets catch the birds with similar effectiveness. This is especially true for special habitats such as desert areas or in spring reeds, where brownish nets are most effective (Figure 2.3). Birds are able to be caught not because of their poor visibility, but because the birds inhabiting reedbeds or bushes do not "know" that they can be caught, i.e. they think "if I can fly through dense reeds or through dense leaves, I can fly through this strange net too". It can be commonly observed that a bird "attacks" the net several times if not caught at the first "trial". So, do not worry if your nets become lighter.

Figure 2.3-1: Nets of brownish colour could be less visible in some habitats than the black ones. Open net. Burullus, Egypt. Photo P. Busse.

Figure 2.3-2: Nets of brownish colour could be less visible in some habitats than the black ones. The net closed. Burullus, Egypt. Photo P. Busse.

2.1.1 Net Poles and Fixing Strings

To erect single net, two poles (for a standard 2 or 2.5 m high nets – 3-4 m long poles are necessary) erected vertically and four strings are needed (Figure 2.1). Poles should be as smooth as possible to simplify erecting and to avoid entangling the netting. Metal or bamboo poles are the best, but they are more likely to be stolen than ordinary wooden poles. Smooth and even slim poles are of a special importance if closing the nets every day is a part of the catching routine. This is a general rule when working in hot climates, i.e. when the midday temperatures exceed frequently 30-35 °C and it is dangerous for the birds to be caught at this time. Fortunately, even in remote areas, there are usually accessible (and not too expensive) aluminium or iron, covered with plastic, tubes used for windows curtains (Figure 2.4-1). Sometimes, when special catching routines are used, poles much longer than normal may be needed (Figure 2.4-2). The net ears, at the ends, are put on the pole, in proper order, and the first pole should be fixed; the net shelves should be still not open – see left person at Figure 2.5-1. Then, stretch the net, still collapsed, along the prepared place. Be careful to not entangle the net in bushes or with foliage on the ground (Figure 2.5-2) – and put nets ears on the second pole. Finally, open the net in such a way that the vertical string is stretched. Erecting two or more nets in a row is a little bit more complicated as the nets' ears must be put on the common pole one by one from each neighbouring net. The strings at every end of the single net (or nets') row must be stretched at angle that protects the nets from falling down when the wind changes direction (Figure 2.6). The distal end of the string should be fixed to the ground by a strong peg or to twigs of bushes and trees. On sandy, soft, and/or wet soils, pegs must be long enough. For some hard, desert surfaces, pegs made from building steel wire (10-12 mm in diameter and 30-40 cm long) are the best. Knots made on the pole should be easy to untie allowing stretching of the net when it becomes longer after some time. The two-tailed pre-prepared strings (shown in Figure 2.7) are very convenient. Whether it is planned to pass under or over the string while checking the nets, the poles can be fixed at different heights. For instance, if we pass under the string, we will fix it higher and fix the other end farther; conversely, if you will go around the string, fix it lower at the pole and closer to the net. However, if we plan to close the nets every day, both ends of the strings must be at the same height, ideally towards the middle of the pole.

Recently, telescoping lightweight poles made of fiberglass became available. They are very easy to transport and operate, including folding to remove birds from the highest shelve. However, telescoping poles are very expensive and attractive to people moving around, so that is why they are not very popular among bird ringers.

Figure 2.4-1: Practical light-weight metal poles – 3-4 meter long (Egypt). Photo I. Rząd.

Figure 2.4-2: Special high nets for catching birds landing on roosting place (Israel). Photo P. Busse.

Figure 2.5-1: Setting the net. First stage. Wadi Allaqi, Egypt. Photo I. Rząd.

Figure 2.5-2: Setting the net. Second stage. Azraq, Jordan. Photo P. Busse.

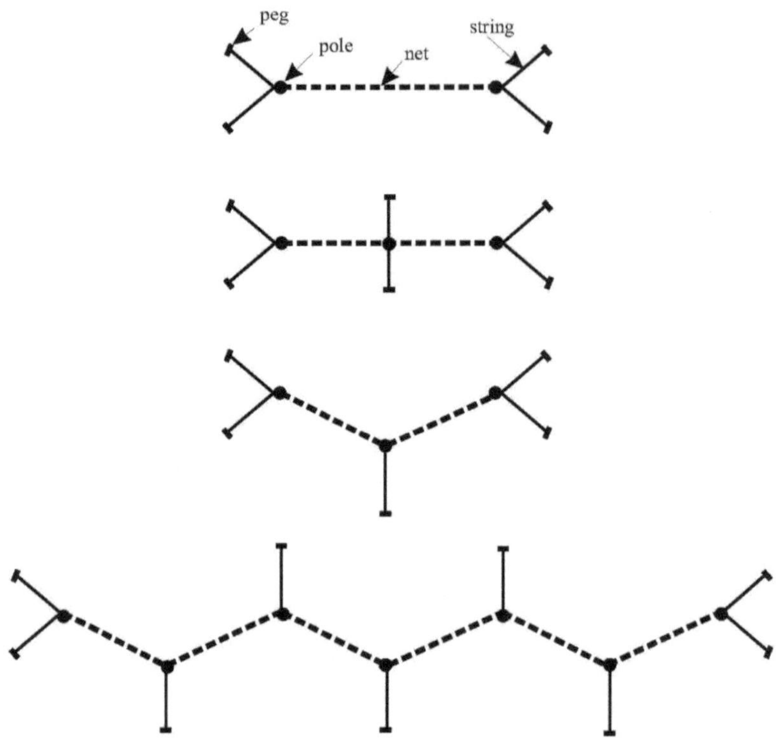

Figure 2.6: Setting up the nets: single net and two (and more) nets in line.

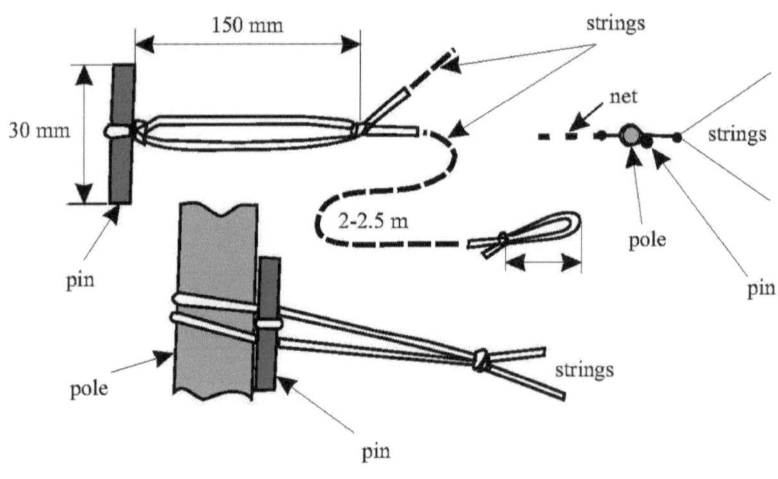

Figure 2.7: Pre-prepared string with pin to fix the net.

2.2 Heligoland Traps

The other important catching device is the Heligoland trap. Originally, it was constructed on the German island Heligoland. The general layout of the trap is a funnel made of net ending with a box or a collecting room from where caught birds are removed. Constructions are permanent or temporary, and their size varies to a large extent. Original traps made on Heligoland are small (around 3 m high) while the biggest Rybatchy-type ones reach heights of 20 m. Dependent on the place where the trap is situated and bird behaviour, there are two types of Heligoland traps in use. One, the so-called "active" trap, is usually a small device situated where diurnal migration does not occur. In that case, the ringer must be active and flush birds feeding or resting in the bushes into the trap. The bushes in front of such a Heligoland trap should be attractive to resting birds and offer good feeding possibilities (e.g. berries; elder, rowan etc.). A pool with drinking water is frequently placed at the front of the trap as an additional attraction. In order to prevent flushed birds from turning around and escaping by way of the entrance, this type of trap is usually constructed in a semi-crescent form. The bulk of the birds caught in such a trap will be nocturnal migrants, so this type will serve as a substitute to nets, which are the best catching devices for nocturnal migrants. The second type of Heligoland trap, the "passive" trap, is located at sites where strong diurnal migration occurs. Birds migrating at low altitudes will enter the trap on their own and, without any flushing, tend to move to the end room. The role of the ringer is limited to removing them from the trap. "Passive" traps are constructed as straight funnels since the birds seldom reverse their direction of movement. Heligoland traps of this kind are very efficient for catching some species also easily caught with mist-nets (e.g. goldcrests, tits) as well as others less frequently caught in nets in big numbers (e.g. Chaffinch, Siskin). Heligoland traps are expensive and vulnerable to strong winds, but in some places they are the best catching device for permanent work. Big Heligoland traps work as a basic catching device in coastal regions of Russia, Estonia, Latvia and Lithuania, in the interior of Ukraine and in Kazakhstan on mountain passes. A particular design of Heligoland trap, a so called "zigzag trap" is used at the Ventes Ragas station in Lithuania. Movable big traps were used at the Operation Baltic stations in the 1960s as a supplement to nets. A few technical details concerning the construction of big Heligoland traps are given below.

2.2.1 Operation Baltic Transportable Heligoland Trap

Since the Operation Baltic stations were temporary camps, a special type of transportable Heligoland trap was in use (Figure 2.8 – after Busse, 1965). The netting funnel, up to 12 m high, 20 m wide and 40 m long, was made from a nylon netting of 15 mm mesh in a front part and 10-12 mm mesh at the terminal part of the trap (Figure 2.9-2). All net-cover was divided into several segments that were stretched by crosswise nylon strings.

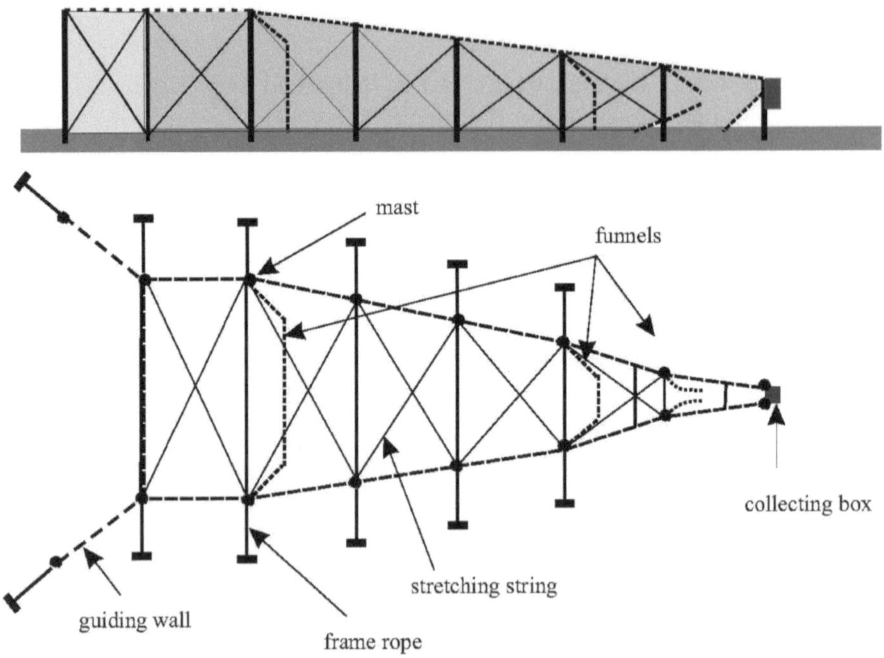

Figure 2.8: Operation Baltic transportable Heligoland trap – side and top views.

Figure 2.9-1: Entrance of the Operation Baltic Heligoland trap at Mierzeja Wiślana, Poland (a view from above of the ending part). Photo P. Busse.

Figure 2.9-2: General view to Ventes Ragas, Lithuania, Heligoland trap. Photo P. Busse.

Three short-funnels made the backwards movement of birds less probable. The construction was hoisted on several metal tubes and fixed by ropes. At the end, the collecting box had different constructs in subsequent versions of the trap, since the first design with glass was dangerous to birds (Figure 2.10).

Figure 2.10-1: One of tried solutions of collecting box in Operation Baltic Heligoland trap. Photo P. Busse.

Figure 2.10-2: In final part of the Operation Baltic Heligoland trap. Photo P. Busse.

2.2.2 The Rybatchy-type Trap

The Rybatchy-type trap is named after the village Rybatchy on the Courland Spit in the SE corner of the Baltic (formerly Rossitten on the Kurische Nehrung, now in the Kaliningrad region, Russia). The trap was designed and has been used since 1957 at the Biological Station Rybatchy of the Zoological Institute, Russian Academy of Sciences. The idea and the construction were due to Jan Jakshis, while Lev Belopolsky and Veino Erik took an active part in the realization of the project. The Rybatchy-type trap was widely distributed on the territory of former USSR in studies of bird migration by means of trapping and subsequent visual inspection of live birds caught.

Between 1957 and 1995, a total of nearly two million birds of 179 species were caught and ringed by the staff of the Biological Station Rybatchy, mainly at a permanent field station "Fringilla", 12 km south of Rybatchy. Up to 1996, these birds have been recovered approximately 7000 times on the migration routes and winter quarters, and there are over 20 000 recaptures recorded at the place of ringing.

The Rybatchy-type of trap does not only catch passerines, but also owls, diurnal raptors, woodpeckers, cuckoos, etc. The highest-recorded trapping in one day at the

Courland Spit was about 9 000 birds in only three traps and 13 000 birds were caught in Kazakhstan in one day.

A preliminary sketch of the Rybatchy-type trap can be found in Belopolsky *et al.*, 1959. Earlier, the trap was described more in detail in Russian publications only (Erik, 1967; Dolnik & Payevsky, 1976), later in the *Bird Station Manual* (Busse, 2000). Although the Rybatchy-type trap evolved from the Heligoland trap, it differs fundamentally from the latter with three distinguishing features:

1. Very large size, with the operating height at the level of bird migratory flight (when the birds fly at a low altitude above the ground);
2. Absence of solid rigid frame - hoisting of the trap by use of steel wires, allowing the possibility to lower the trap before an approaching storm in order to protect the netting;
3. No food or water is used to attract the birds, so the trap can be established in any area with intensive bird migration, even in a desert.

So, the Rybatchy-type trap is basically a huge funnel made from a thread net fixed to the ground and opened towards the stream of migrating birds. In most cases the birds themselves (without particular flushing) reach the terminal part of the trap, the so-called "collecting box", from which they cannot leave.

Construction of the trap is comprised of four basic elements: frame, netting, operating devices, and collecting chamber. The foundation for the trap is made up of four pairs of pillars (Figure 2.11). These pillars are fixed in position with steel wires. The overall length of the carcass construction is 60 to 80 m. The front (first) pillars may range in height from 12 to 15 m, while the heights of consecutive pillars are 7, 4 and 2 m. The distance between the pillars of the first pair (the width of a gateway) is 30 m. Distances between the pillars of following pairs are 15, 7 and 2 m, respectively. The distance between the first and second pair of the pillars is 30-40 m, between the second and third pair is 15-20 m, and between third and fourth pair is 10-15 m. At the gateway, guiding walls may be used. An additional pair of pillars is necessary for these additional walls. All the pillars must be fixed not only by the upper carcass wire, but also by two stretching steel wires. One end of the wire should be fixed to the top of the pillar and the other one to the ground. Tension may be obtained from screw coupling. Wire diameters of 8-10 mm serve well. The pillars may be wooden as well as different material, e.g. reinforced concrete, open-work metallic construction, metallic pipes, etc. The wood pillars, especially with concrete "feet", have some advantages over other materials: they can be mounted vertically without a crane, for example with the help of winches. However, the wood pillars are short-lived, if not impregnated enough, and must be replaced after five years of service.

The whole thread net trap should be made as one unit (walls and ceiling), separate from the carcass. The size of the netting should be smaller by 1-1.5 m than the size of carcass. Different types of netting may be used, cotton as well as synthetic thread. The cotton nets are more durable. UV rays could destroy the synthetic thread within

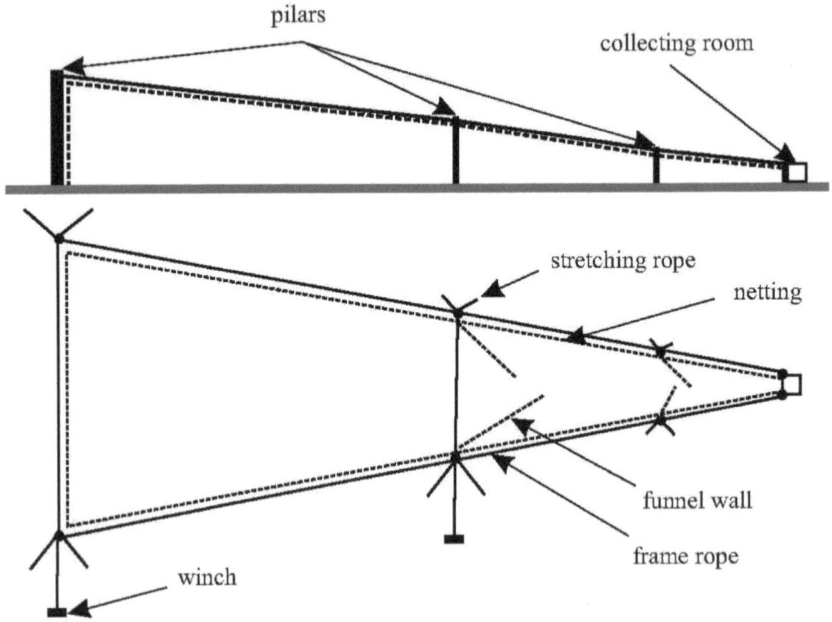

Figure 2.11: Rybatchy trap: side and top views.

a few months. However, netting made of synthetic thread may be more useful when it is necessary to reduce the weight of the trap and counteract a detrimental effect of strong wind in open country. Different mesh sizes are used in different parts of the trap. For the ceiling in the front of the trap, 30-40 mm (knot to knot) mesh net is used. Narrow strips of 16 mm net are used on both sides of the ceiling. Walls of the front part and the whole middle part (from the second to the third pillar) are made of 12-16 mm mesh net. The rest of the trap is made of 8 mm mesh netting. The durability of the trap will be prolonged if there are 5 mm cords attached to the net at 5-6 m intervals along the length and crosswise to reinforce subsequent segments. Inside the trap, it is necessary to make two pairs of so-called "false walls". The false walls are made from the same type of net as the main walls. Their purpose is to keep birds from changing direction of flight. After passing the first false walls, it is more difficult for birds to turn around than to continue to the terminal part of the trap.

Devices for hoisting and closing the trap include metallic rings of diameter 30-40 mm made from wire of diameter no less than 4 mm, which should be attached to the cross cord located at the level of the second pair of pillars. Steel wires of diameter 8-10 mm hoist the trap and pass through these rings before going on to the winches. It is possible to hoist the trap by two cross ropes and two winches only. Different winches may be used. At the Biological Station Rybatchy, the big stationary winches with carrying capacity 1.5 tons have been in use for 30 years. It is also possible to use little winches fixed on those pillars that to where the ropes are also attached.

Variations of the **terminal part** of the Rybatchy-type trap. The Rybatchy-type trap can be adapted according to local conditions (localization, financial possibilities, number of persons in the staff, etc.). The differences primarily concern the final part of the trap. In its initial form, the final part of the trap is arranged in the following way (Figure 2.12): The last 10-15 m of the trap is a narrowed corridor approximately 2-3 m wide and 2 m high. The corridor ends by the sloping wood sheet, which directs the birds into the open cone. At the height of 1.5 m above the ground, the cone leads into the collecting room or collecting box. It is possible to have a system of two cones inserted one in another. The collecting room is made in the form of a netted box with approximate size 1.8 × 1.3 × 1.3 m. The person inside the box may take birds by hand. A second room of the same size is connected to the collecting room, and it prevents birds from escaping when a person is entering the collecting room. The trap may have up to three collecting rooms with permanent or movable cones. Other solutions are used elsewhere with this type of traps (e.g. in Ventes Ragas big Heligoland trap – Figure 2.13-1).

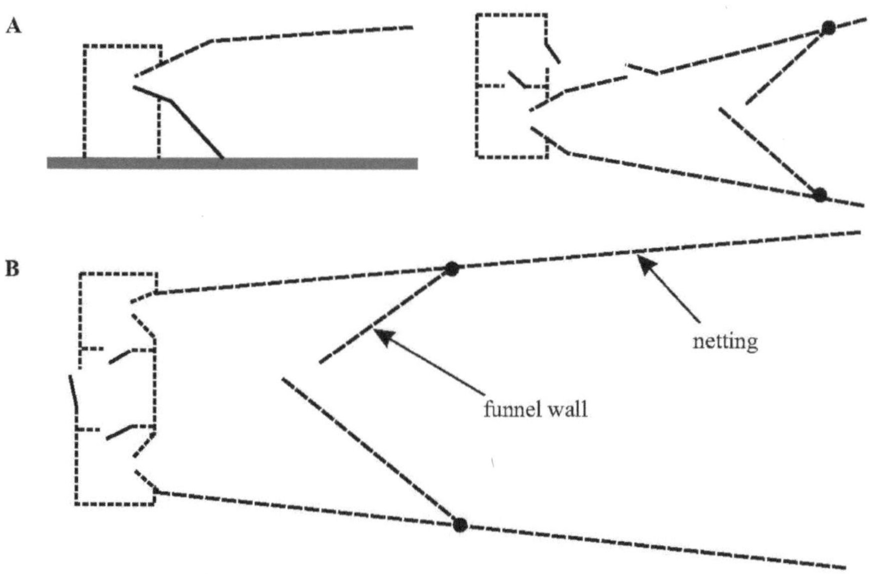

Figure 2.12: Terminal part of the Rybatchy trap. A. One collecting room (side and top views), B. Double collecting room (top view).

Figure 2.13-1: Final chamber in Ventes Ragas, Lithuania, big Heligoland trap. Photo P. Busse.

Collecting boxes could be even removable and the staff changes the box with birds against an empty one. Such boxes may be made of net or transparent plastic (Figure 2.13-2). They are used in different types of Heligoland traps and sometimes also in the Rybatchy-type trap. For example, at the Ladoga Ornithological Station (village Gumbaritsy at Lake Ladoga) the Rybatchy-type trap has a small removable collecting box made of a wire frame covered with netting. Such boxes are mounted on the terminal cone of the trap. In Kazakhstan, at the Chokpak ornithological station, the trap has a collecting box similar to the one used at the Swedish ornithological station, Ottenby. Birds entering the end cone of the trap fly towards the transparent window, strike upon it and slide down into the small box (Gavrilov, 1968).

A collecting room has some advantages compared to the small collecting boxes. During intensive bird migration, the collecting box will at times get filled with birds in an instant, and it may happen that passerines and raptors are indiscriminately mixed. The larger volume of a collecting room may save the small birds in such cases. When the birds have been removed from the collecting room, they are put into special portable boxes. These have low walls and a small mesh netting top.

Figure 2.13-2: Transport of the birds caught at Ventes Ragas big Heligoland trap. Photo P. Busse.

2.2.3 Zigzag Trap

The zigzag trap is a new type of trap for bird catching, based on the idea of Heligoland trap; however, to differentiate the two, the zigzag trap allows birds to move in two opposite directions. This is a unique feature in trap design. The trap is mainly designated for catching passerine birds. However, diurnal and nocturnal raptors (mostly Sparrowhawks), Cuckoos, woodpeckers, and other birds are also found in the trap quite frequently. This trap may be used to catch birds under every possible weather condition. The only danger to the trap itself can be heavy snow, but the main benefit is that captured birds are not entangled in nets and not injured. L. Jezerskas, the former head of Ventes Ragas Ornithological Station in Lithuania, constructed the trap. Three traps of this type were built in the years 1982-1984. Jezerskas (1990) has described the construction of the zigzag trap. A total of 162 944 birds of 128 species were caught using these three zigzag traps at Ventes Ragas Ornithological Station in five years (1985-1989).

In principle, the zigzag trap is a system of modified Heligoland traps, connected sideways with their gateways directed in opposite directions (Figure 2.14 – after *Bird Station Manual* - Busse, 2000). The size of the trap depends on the number of the sections and their size. The number of the sections is unlimited in one trap. It can be as large as the confines of a given site allows. The size of the sections can be different in different traps, but it is recommended that one trap contains one-size sections. In the opposite end of the entrance, every section has a bird collecting chamber and a basket. The chamber is shielded with a "roof". The "roof" protects the birds in the chamber from direct sunrays and rain. The top ("ceiling") of each section up to the middle is horizontal, from this point it gradually ascends to the beginning of the collecting basket. Recommended dimensions of the sections are as follows: the length (up to the beginning of the basket): 12.5 m, the width of the front: 15 m, and the height of the front – 6 m.

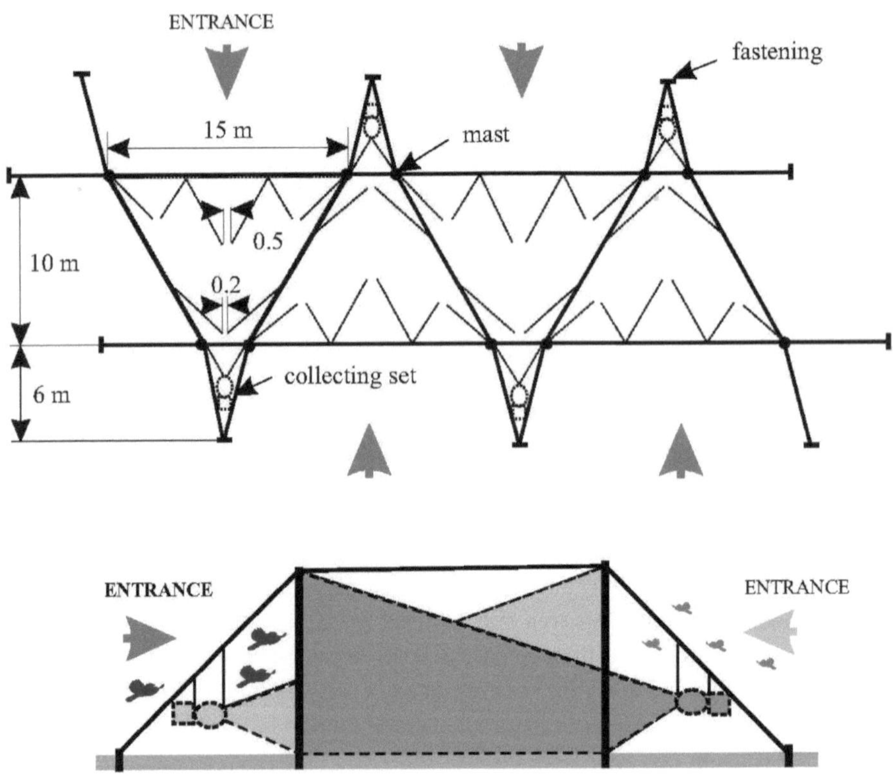

Figure 2.14: Zigzag trap - top and side views.

There are three entering vertical slits, each 0.5 m of width in the front part of each section of the trap. The length of the slits is equal to the height of the trap. The first pair of "wings" forms these entering slits. There is a second pair of "wings" in the further interior part of each section of the trap. The width of slit between the second "wings" pair is 20-25 cm. The "wings" prevent birds from getting out of the trap.

The materials necessary for the arrangement of the two section zigzag trap are listed below (one section is not a zigzag trap yet!). In parentheses, the amount of the materials necessary for each additional section is given:

Metal pivots (diameter 25-30 mm, length 0.8 m) – 8 (2);
Metal plates (thickness 15-20 mm, size 20 × 20 cm) – 8 (2);
Metal pipes (diameter 60-80 mm, length 6 m) – 8 (2);
Metal fastening hooks (made of pivot diameter 15-20 mm, length 0.8-1.2 m depends upon hardness of a ground) – 8 (1);
Steel rope (diameter 6-8 mm,) – around 155 (55) m;
Metal wire (diameter 6-8 mm) – around 24 (12) m;
Nylon string (diameter 6-8 mm) – around 165 (60) m;
Nylon string (diameter 4 mm) – around 165 (60 m);
Nylon net (mesh size "knot to knot" 14-16 mm) around 980 (450) m^2;
Nylon net (mesh size "knot to knot" 8 mm) around 24 (12) m^2.

Firstly, the frame of the trap must be arranged. The metal pivots are beaten vertically into the ground at the points shown as dots at Figure 2.14. Around 20 cm of the pivots are left above the surface (it is recommended to paint the pipes in pale colours). A hole is drilled in the centre of the metal plate. The diameter of the hole has to be around 2 mm larger than the diameter of the metal pivots beaten into the ground. These metal plates are pulled on the pivots. The masts (made of metal pipes) are put onto the ends of the pivots left over the surface of the plates. The metal plates prevent the pipes from going into the ground. In upper part of the masts (5-8 cm from the top) there holes of 8-10 mm in diameter have to be drilled. Masts are connected by the steel rope, which passes through the holes in upper parts of the pipes. The loose ends of the rope are strained and fastened to the metal fastening hooks, which are beaten into the ground. The masts fastened in this way must remain in a straight vertical position and must not move.

When the frame of the trap is ready, it is time to make a trap itself from the 14-16 mm "knot to knot" mesh nylon net. The net is cut into appropriate pieces of the necessary size, which are sewn together with a thin nylon string. The 6-8 mm nylon string is fastened (sewed) in the place where the top ("ceiling") and the sides ("walls") come together. The same kind of string is fastened (sewed) to the top ("ceiling") front edge of the trap, on the bottom and on the front edge, which reaches the metal pipe of the sides ("walls"). The 4 mm nylon string is sewn to the edges of the first and second pairs of "wings" and at the intersection of "wings" with top ("ceiling") and sides ("walls"). The trap is fastened by 6-8 mm nylon strings to the metal pipes and, if necessary, to the steel ropes connecting the metal pipes. The bottom part of the first and the second pairs of "wings" and the bottom part of the sides "walls" are fastened to the ground.

Bird collecting basket and a chamber are made at the end of each section (Figure 2.15). Their frames are made of 6-8 mm metal wire and covered with 8 mm "knot to knot" mesh nylon net. The 50 cm length "sleeves" made of the same kind of net are sewed on the side of chamber and basket. The "sleeves" are used for removing the birds from the chamber and the basket. After removing the birds, the "sleeves" are tied. The basket has the form of an egg, and the more pointed- end is directed towards the chamber. Its frame consists of 8 low-shaped longitudinal and the 3 circle-shaped wires connected to them: one (diameter 20 cm) at the end, another (diameter 40 cm) in the opposite end and the last one (about 60 cm in diameter) in between. At the end where the basket joins the trap, the diameter of the circle is 40 cm. The chamber has the form of a cube, the edge of which is 50 cm long. The chamber and the basket are joined together at the sharp end of the basket. There are two downward "gullets" fastened to the each end of the basket. The diameter of the narrower end of the both "gullets" is 10 cm. The diameter of wider end of the same "gullets" corresponds with the diameters of the circle-shaped wires in the ends of the basket (20 and 40 cm). The bigger "gullet" is around 35 cm long and is pointed inside the basket while the smaller one is about 20 cm long and is pointed outside the basket, i.e. inside the chamber. The frames of the both "gullets" are constructed of 3-4 mm wire and covered with 8 mm mesh size nylon net. The baskets and the chambers are hung on 6-8 mm metal wire on the steel ropes fastened to the metal hooks beaten into the ground. Chambers have roofs that are made of reeds, tarpaulin or other materials.

As the catching season comes to an end, the baskets and the chambers are removed. The trap and the frame are untied and the trap is also taken off. The trap is stored indoors till next season.

Zigzag traps have some advantages for work at regular, year-round ringing stations and can substitute mist-nets as well as traditional big Heligoland traps:

- First of all, they can work in spring and autumn as the same construction set in a field stand – they could catch the birds moving back and forth. That is very important in places where there is no defined and narrow stream of migrants, but a more or less uniform habitat (e.g. reeds and bushes) where migrants stopover during migration.
- The process of removing the birds from the zigzag trap is easy, short, and uncomplicated. It can be done even by less experienced staff or by helpers. Conversely, the process of removing birds entangled in mist-nets, especially the removal of tits is time-consuming, tiring and demands a lot of manpower and competence; they are among the birds most commonly caught at many stations.
- In the zigzag trap, the birds are less exposed to adverse weather factors since they have ample space to move in the chamber or in the basket (there is a roof on the chamber that protects the birds from direct sun rays and rain).

- It is almost impossible to overlook a bird in a zigzag trap.
- In zigzag trap there is no problem with leaves, twigs, bigger insects etc. that is so well known when using mist-nets.
- Zigzag traps are efficient under all meteorological conditions.
- Closing and re-opening of the zigzag trap is extremely easy: one only has to close or open the first "wings".

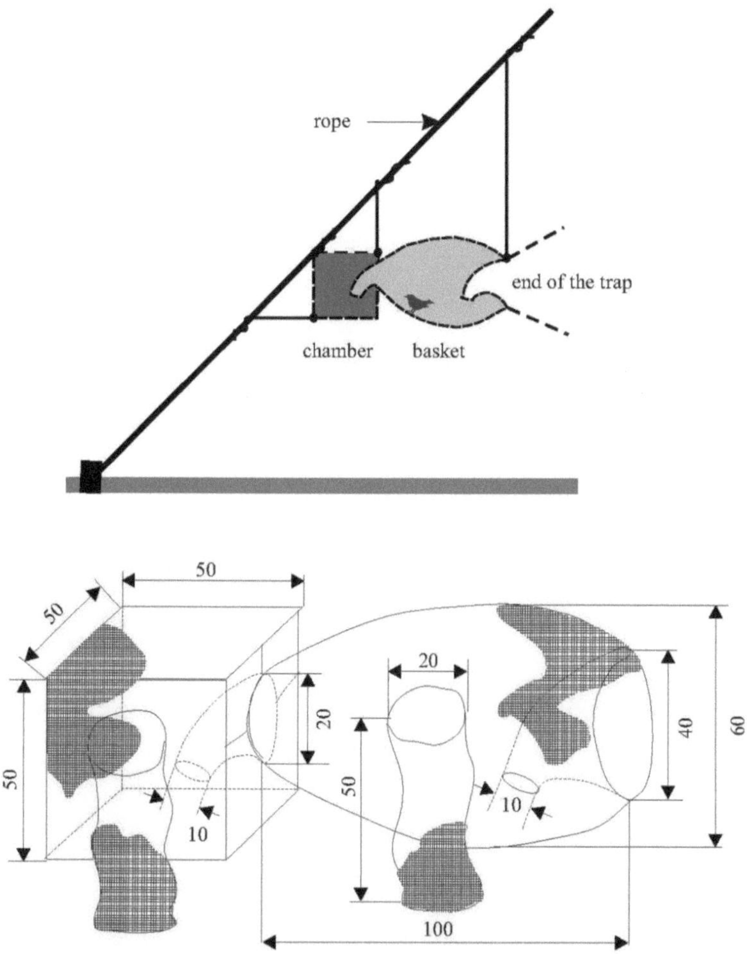

Figure 2.15: Zigzag trap – terminal part side view (upper), collecting basket and collecting chamber (measurements in centimetres).

2.3 Funnel Traps

Wader funnel traps are very specialized catching devices and they are presented in *Walk-in Traps* section (p. 130). They are efficient for catching some ground-feeding passerines as well, e.g. wagtails, pipits and starlings foraging on beaches and meadows.

3 Passerine Station Laboratory Equipment

The basic laboratory equipment should allow ringing and collection of standard measurements in an effective way and with the smallest possible effort. To some extent, the working routines at the laboratory site decide the needs, and not all items are necessary at all sites. The laboratory stand and tools presented here were carefully elaborated from an ergonomic point of view and checked during 50 years of Operation Baltic.

3.1 Bird Transport and Storage Devices

The basic container for transportation of passerines to a laboratory is a linen bag closed with a soft string (Figure 3.1) that could be hung up on a special hanger at the chest of the ringer (Figure 3.2), on a binocular, which is a very convenient solution, if the person is not simultaneously the passing birds observer, or, in the worst case, on a forearm (Figure 3.3), but not carried in the hand. Generally, two kinds of bag strings could be used: without or with a lock. In the first case, the string could be slippery, of a synthetic fabric. In the second case, string must be not too smooth, synthetic, but a cotton one. Both of these constructions have their own pros and cons. Bags with locks protect us better from escaping birds and make the sporadic addition of new birds easier, but the strings can easily entangle with each other when not well-packed into

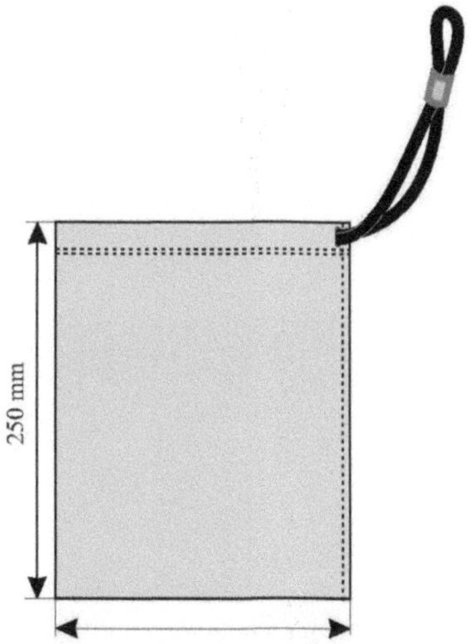

Figure 3.1 Bag for bird transportation.

Figure 3.2 Breast hanger for bags.

Figure 3.3-1: Extracting the owl during night net control. Kopań, Poland. Photo unknown.

Figure 3.3-2: Good catch of owls. Because of weight of the owls caught bags are hanged on forearms. Kopań, Poland. Photo W. Busse.

the pocket. The bags with strings without lock are more resistant to entangling, but if they are not closed properly, birds can escape. Additionally, if bags are closed, as many helpers do, the strings can get knotted, and these knots are not easy to remove, especially when wet. The size of the bags may differ – a small “one bird-person” bag may be used for transporting single individuals such as rare birds or birds with a foreign ring, while standard bags (approximately 20 × 25 cm) may be used for the majority of birds, but a different number of individuals according to size (Table 3.1), and special bags for bigger birds (e.g. owls, raptors, small waders). Standard bags should be numerous enough, at least 100 at the station where large number of birds is expected, since they are used to transport birds from the nets and temporarily store them while they are waiting for ringing and investigation (Figure 3.4). A deficit of free bags sometimes may cause dramatic disturbances in station work during peak days; birds will suffocate in overcrowded bags. Because of this, at the laboratory, the birds should be stored (when numerous) in special storing devices like boxes or baskets, where they have more space and do not risk suffocating when wet. One type of bags in use has hard plastic bottoms. Such bottoms should be with many holes, e.g. made of hard plastic netting, allowing excrements to drop out; in the worst case the birds will get dirty and wet, and the ringer is the cause of their impaired condition.

Table 3.1: Number of individuals allowed to be transported in one standard bag (20 × 25 cm).

Species	Number of individuals*
Goldcrest	10 - 15 – 20
Long-tailed Tit	8 - 10 – 15
Blue Tit, Coal Tit	7 - 10 – 15
Great Tit, Robin	6 - 8 – 12
thrushes	1 - 1 – 2
Jay	1-1 – 1

* First number indicates when birds are wet or they must wait longer time; Central number indicates standard number of individuals; Last number indicates number of individuals when they will be immediately put into storing devices.

The most important thing when preparing bags is to use linen that allows for the passage of air. Cotton bags usually have been washed before first use for removal of chemical apertures. A disadvantage of cotton bags is that they easily absorb water from excrement and moist birds. It takes long time for them to dry and they are more likely to be damaged by microorganisms when moist. Synthetic linens are much more excrement resistant, but they must have visible holes between threads, to be air transparent enough. Wet birds stored in synthetic bags are still wet when you remove them. On the other hand, wet bags could be easily dried. Bags should be regularly washed and in the meantime cleaned from droppings and feathers.

Figure 3.4-1: Good catch of roosting birds. Kopań, Poland. Photo W. Busse.

Figure 3.4-2: Bags with the birds caught at the roosting place. Kopań, Poland. Photo W. Busse.

It is not advisable to use storing boxes or baskets (see below) to transport birds from the nets. There are only few exceptions when this is acceptable, e.g. when mass catches of one species occurs in a limited number of nets situated close to each other. In such a case one particular person, besides the person making regular checks, should use them, but birds should be placed in a bag first and then shifted into the box. The birds caught in Heligoland traps may be transported in the final trapping boxes if such are included in the construction of the trap.

During peak days, when a few species tend to be very numerous, it is convenient, sometimes even necessary, to use bird-storing devices where birds could wait for ringing. Such devices can be made from different boxes, baskets etc. (Figures 3.5 and 3.6-2). The most important points of construction are:

1. Free access of air – at least part of the walls must be made from small mesh netting. The meshes must be smaller than any bird head (heads must not go through!);
2. Easy handling of birds: they should be easily put into and in particular easily removed from boxes at the ringing stand (boxes are used when you are in a hurry!),
3. Easy to move – storing devices should be kept in a cool, dark place and then moved to the ringing stand. But note: movable does not mean "used at the nets"; this is acceptable only under special conditions (see above).

The holding capacity of any particular size of box or bag depends on the particular species that is going to be stored in it, this must be estimated from case to case. Each bird should be able to sit on the floor; in most species the limit is set by that area of the bottom. Some species, however, will cling to the bag walls or to the ceiling of a box (e.g. goldcrests, tits), and the number may be increased accordingly. But watch out for indoor temperature increases when there is much stress among the birds; a bag full of "over-heated" goldcrests or siskins will kill itself in no time, and the losses of humidity may be harmful to birds.

Figure 3.5: Example of how Passerine birds are stored.

Figure 3.6-1: Good catch of tits in the net. Kopań, Poland. Photo W. Busse.

Figure 3.6-2: Many tits waiting for ringing. Kopań, Poland. Photo W. Busse.

It is better to have few smaller boxes than a few big ones. In one box, birds from only one check, and obviously of one species, can be stored.

In exceptional cases, boxes may be used for overnight storage of diurnal migrants ringed late in the evening when **the weather is bad** (rain, fog, snowfall) or the owls migration is ongoing. In such cases, the number of birds per box should be reduced by more than 50 percent of the standard. Night migrants should be released during the night unless many owls hunt around. At roosts, swallows and wagtails may be let free in total darkness when they are no longer blinded; they settle in the reeds without delay. If bigger number of birds are ringed and released in darkness, an additional check of nets close to the ringing site should be performed.

3.2 Laboratory Tools

The ringing laboratory tools are shown in Figure 3.7.

Ruler. Depending on method of measurement, different types of rulers are used (Figure 3.7 (1-3)). The most common one is a metal ruler with a stop at zero-end. It is used for a maximum chord measurement of the wing, while tail measurements are not possible with this type because of the stop. Wing-formulas may be measured with this ruler as well, but the procedure is less convenient. Stop rulers must be carefully checked for precision. In some cases, the stop is not properly fixed. Special rulers with a pin (1.4 mm in diameter) fixed perpendicularly to the ruler at zero line is used for 3rd primary measurement. The most universal tool is, however, a ruler without the stop and cut off exactly at the zero-end. The length of this ruler is 30 cm, and it can be used for wing-formula, wing-length and tail-length measurements. Wing-length measurements taken with this type of ruler are exactly comparable with measurements taken with the stop ruler and a common opinion about their lower accuracy is unsound.

Balance. Very convenient, and not too expensive, are electronic balances with digital reading and a battery power supply ((Figure 3.7 (5)). Loads up to 500 g and exactness 0.1 g are ideal for most of netted passerine birds. There are balances of even more than 500 g capacity and 0.1 g exactitude. The balance of bigger capacity and lower exactness could be useful when waders or raptors are caught more frequently. The birds are weighed being put on the balance inside of a conic plastic tube adjusted to the size of the bird; there will be a resulting pull of 1 – 5 g when the weighing in a bag and the bird flutters from side to side. The balance should be protected against wind that can greatly disturb reading. The most common type of balances used in the field by amateur ringers is spring balances of Pesola type (Figure 3.7 (4)). They are intended for different sizes of birds and measure with different degrees of precision. A full-scale load of 30 g and exactitude 0.1 g is applicable for most small passerines.

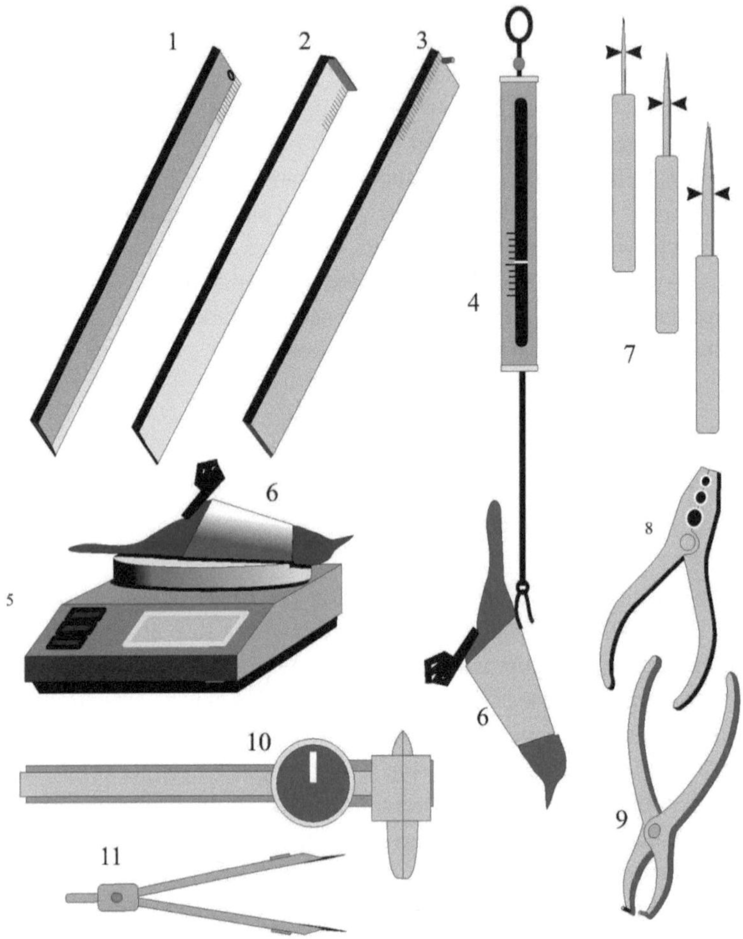

Figure 3.7: Laboratory tools. 1. Ruler without a stop, 2. Ruler with a stop, 3. Ruler with a pin, 4. Pesola balance, 5. Electronic balance, 6. Tube for weighing the birds, 7. Openers for opening rings (note that from arrows to the hand side surfaces must be parallel), 8. Pincers for closing rings, 9. Reverse pincers for opening rings, 10. Callipers, 11. Dividers.

Balances of bigger capacity (full scale 100 g, 300 g, 1000 g) and lower exactitude come into use where heavier species are involved. For small birds, their exactitude is inadequate. The birds weighed are hanged to the balance in a conic plastic tube (Figures 3.7 (6) and 3.8) adjusted to the size of the bird (weighing the birds in bags is not recommended). For the modern station work Pesola balances could not be recommended because of two properties: it is much more time consuming, especially if one has a bad habit of weighing birds in bags, and it loses precision with time. The spring elongates during normal use and also extends at high temperatures while shrinking at cold temperatures, so it should be calibrated at both ends of a scale.

Figure 3.8: Weighting of the bird using Pesola balance in the last century. Mierzeja Wiślana, Poland. Photo Unknown.

3.2.1 The Orientation Tests Equipment

The described method of studying directional preferences in nocturnal migrants includes a new field technique that pays special attention to the inconsistency of directional behaviour pattern in an individual bird. It may be used under real field circumstances, by professionals as well as amateurs. The equipment is simple and cheap, and the technique easy to learn in a standardized form. Additionally, the test routine allows for a great deal of data collection, since tests may be performed during both night and day. Diurnal tests under an overcast sky have the same value as tests done with good sky visibility, which is not the case in nocturnal tests. Analysis of local vectors in a directional behaviour pattern seems to be of use in the studies on local migratory directions and the overall population composition of migrants. The method used commonly by researchers (Emlen's cage) is much more stressing the birds, needs much more effort (time and material). The equipment set includes:

1. Circular, not transparent, uniformly coloured screen, which keeps the bird in an experimental cage from seeing any landmarks, trees, wires etc. (Figure 3.9). Its diameter is 110 cm and its height 40 cm and it is made of four sectors for easier transportation (Figure 3.10-1).

Figure 3.9: Orientation tests set: protecting wall around and experimental cage inside.

2. Testing cage (Figure 3.10-2): a cylinder cage made of two wire circles connected by eight vertical wires distributed evenly. These define sectors used when counting results. Diameter of the cage: 36 cm, height: 10 or 12 cm. Higher cages are used for testing thrushes, but they may be used for smaller birds as well (since differences between results obtained with these two heights were not found). The top surface of the cage is covered with nylon netting of 10 mm mesh. The sidewall is covered by a strip of ultra-thin, transparent plastic foil used to keep food in refrigerators (sold in rolls).
3. Not too slippery plate of neutral colour as a bottom surface under the cage or piece of linen to cover smooth ground.
4. Forms for recording the collected data.
5. Pointed colour marker.

Some useful details of the orientation stand equipment are in the description of the method (*Special Studies*, p. 102).

Other tools. Bird rings may be opened with particularly designed openers (Figure 3.7 (7)); if there are not many to do, a sharp knife or hard nail will suffice.

In many ringing schemes, **pincers** with side holes are used for closing all rings. This gives a very exact closure of the ring, without fissure that might be harmful to the bird. In some countries, small aluminium rings are simply closed with the fingers, while larger rings (with locks) and rings made of stainless steel are closed with pincers (Figure 3.7 (8)). The "finger closing" technique needs very high quality rings, preferably not much opened and not too hard. Closing small rings with the fingers will speed up the ringing procedure, but the ringer must be well trained and take care

Figure 3.10-1: The orientation tests stand in a field. Burullus, Egypt. Photo P. Busse.

Figure 3.10-2: Demonstration of internal orientation cage. Photo P. Busse.

to close the rings properly. However, after a day's hard ringing work, the fingers will be very sore and could be less accurate!

Special technical **reverse pincers** (Figure 3.7 (9)) may be used when a closed ring has to be removed from a bird's leg, but note that operations of this kind can be performed only exceptionally, when the ring is dangerous to the bird. In most cases, removal of a ring is very difficult and the whole procedure may injure the bird leg. When a stainless ring is bent over it may be better not to mess up things more than they already are. However, one ringer had the idea, apparently effective, to use two loops made of strong, but thin stainless steel wire loops with little hands as shown on Figure 3.11.

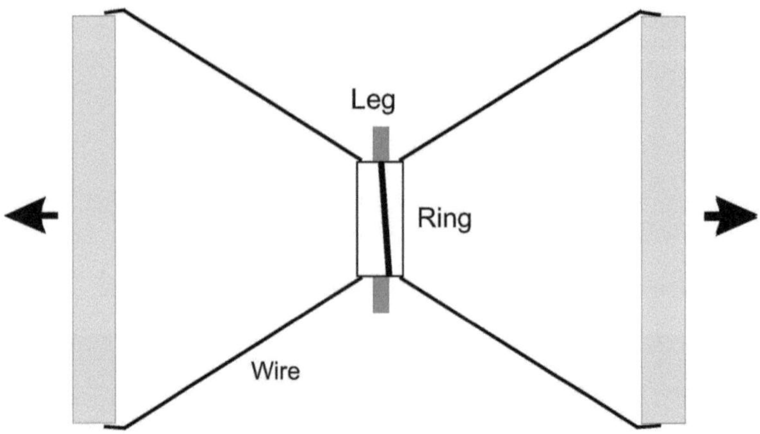

Figure 3.11: Opening of the stainless steel ring.

Colour rings need special applicators to put them onto the bird's leg. There are different applicators depending on the type of plastic rings. Colour plastic rings must be glued when closing.

Callipers (Figure 3.7 (10)) are used for tarsus and bill measurements. They must be of good quality and slip easily - the bird tarsus is not made of iron! The best, but also most expensive, are modern types with digital reading. Less expensive ones are plastic callipers with round, clock-like scale. The cheapest, but definitely the worst quality, are the traditional ones; however, measurement is much more time consuming.

Dividers (Figure 3.7 (11)) are used for tarsus, bill and sometimes tail-length measurements. Measurements taken in this way are less precise and more time consuming. Recently, dividers have gone out of use.

3.3 Rings

The rings must be open and easily removable from sticks or strings. They are supplied in an opened state and either threaded on a plastic string or stored in a tube. However, to remove them from the string could be difficult during ringing work; all rings will not easily slip from the elastic string. So, best way to prepare in advance is to shift them from the string to a metal wire stick of a slightly smaller diameter (Figure 3.12). Rings must be open enough to be put on a tarsus of the largest birds ringed with the particular ring size. If they are too closed, the ringer should open them to proper fissure size in advance and put them on adequate sticks. The rings must be opened only with a proper opener. Rings opened in a wrong way (Figure 3.13) cannot be properly closed and could be harmful to the birds. The bottom side of the ring number must be down the stick. Rings of rarely used sizes may be kept unopened and stored on the original strings. After ringing, its position on the bird leg is most convenient when ring number is read during retrap handling; this greatly reduces number of reading errors. Most small passerines are checked as re-traps in the hand, not by sightings (as for bigger birds), so this style of positioning rings prevents misreading the number, especially in hurry, the ring number is better protected against wear when it is located at the bottom of the ring. However, rings put on waders and birds should be oriented in such a way that the number could be easily read from a distance.

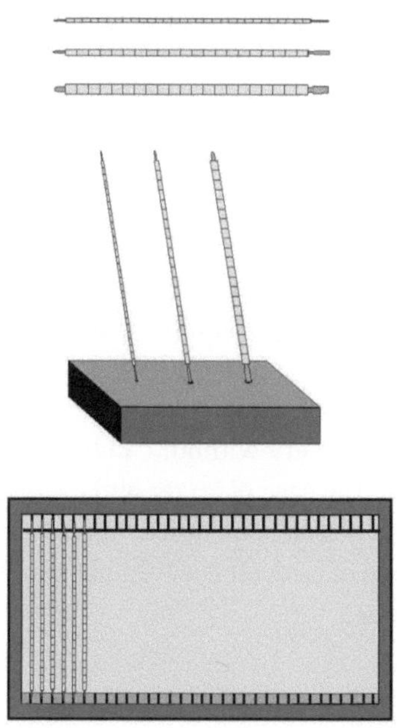

Figure 3.12: Handling of open rings: the sticks for different sizes of rings, working position and a box for storing the sticks.

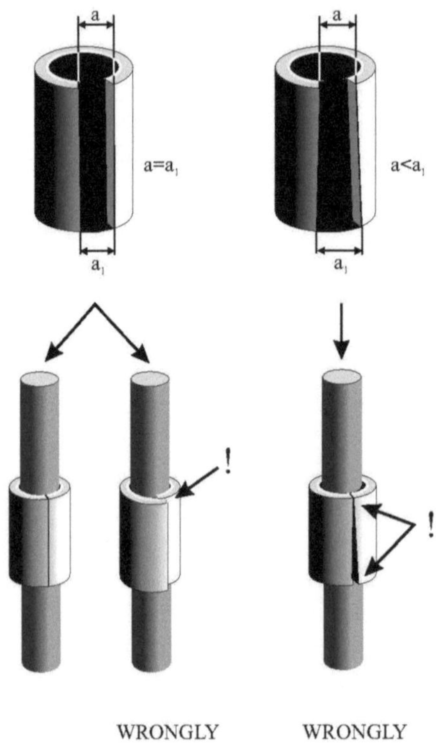

Figure 3.13: Correctly (upper left) and incorrectly opened ring (upper right); correctly (lower left) and incorrectly closed rings (right). Note that incorrectly closed rings are harmful to the bird!

Colour rings are more often used in ringing programmes run during migration (Meissner & Bzoma, 2011). They are also useful in ecological studies during the breeding season. In recent years, the number of different colour ringing schemes has increased (see www.cr-biding.org). The main benefit of using colour rings over standard metal rings is the possibility of obtaining a recovery without catching the bird. Colour rings increase recovery rates, especially in the case of larger birds, when the inscription on the ring is big enough to be read easily from a distance. In projects involving individual marking of hundreds or thousands of birds, it is advisable to use plastic rings engraved with a unique field-readable code.

3.4 Ringing Stand

The ringing stands may be different from that described here according to logistics and expected number of the birds caught, but ultimately, they should be comfortable for people and birds.

The ringing stand (Figure 3.14) should be fitted with two or three comfortable seats, and a table that allows convenient writing in notebooks (see *Laboratory Working Routine* – p. 118); you may have to spend many hours without a break there. A couple of rows of hooks for bags with birds should be fixed to the edge of the table, the distance between each hook should be so large that the bags do not press against each other. Every row of hooks is intended for birds removed during one control walk. It is advisable to have at least 4-5 rows, as there may sometimes be a need for more than one row. One of the rows should be within hand range of the ringer. In front of the ringer, a row of sticks with rings of sizes most commonly used should be placed. The rings MUST be open and easily removable from the stick or string. Consecutive sticks or strings of rings should be easily accessible to the ringer and stored in top order to prevent one of them from being left out when the next sequence is needed. A set of ringing tools should always be placed within hand range: ruler, pincers (used at least for larger rings and rings of stainless steel), callipers or dividers (if used for special measurements) and scissors (accidentally used). The balance should be conveniently located according to the type used and the organization of work (as to who reads the weight: see *Laboratory Working Routine* – p. 118). If the ringing stand is organized in a room or closed tent, a convenient releasing funnel is needed. Passing the bird to another person to release it outside of the room is extremely inconvenient, and frequently birds escape into the room if workers are not trained well in passing birds (see later in *Normal Routine* – p. 119). For evening and night ringing, a good source of light must be arranged. The best are halogen 12 V lamps, while lighting diode head-torches could be useful too; note, however, that this kind of light changes our perception of some colours, thus, e.g. fat scoring is much more difficult as well as discrimination of some species or age and sex groups by colouration.

When the weather is warm enough and not too windy, the best solution is to have the ringing stand in the open (Figures 3.15 and 3.16), but under a tarpaulin roof protecting against rain and direct sun. Only very temporary ringing stands could be unsheltered (Figure 3.17, 3.18-1). In lower latitudes, a shade stand is absolutely necessary. One advantage with the open-air laboratory is that the bird can be let free without delay when it has been handled, and if it escapes, it can fly freely without hitting a window. In an unsheltered laboratory, however, the conditions will sometimes be too windy for accurate weighing, or a little on the cold and wet side for human beings – although nobody died as a consequence of that as far as reported. The material of the roof should be of neutral colour (white or grey) and preferably half-transparent since good light will facilitate sexing and ageing based on subtle colour characters. The same applies to the colour of walls inside a laboratory room.

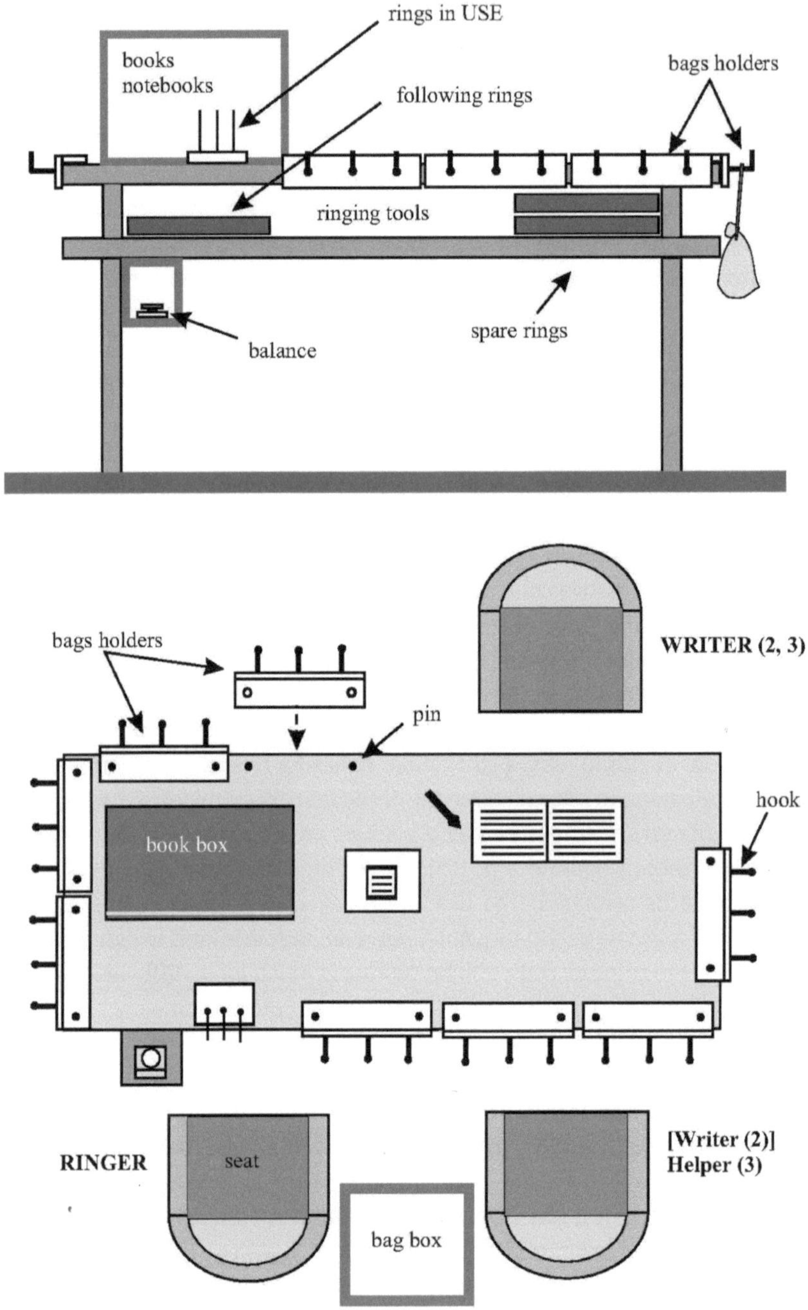

Figure 3.14: Fully equipped laboratory stand: explanations in the text.

Figure 3.15-1: Seasonal bird ringing station - Kopań, Poland. Photo P. Busse.

Figure 3.15-2: Seasonal bird ringing station – Mouth of Vistula, Poland. Photo W. Meissner.

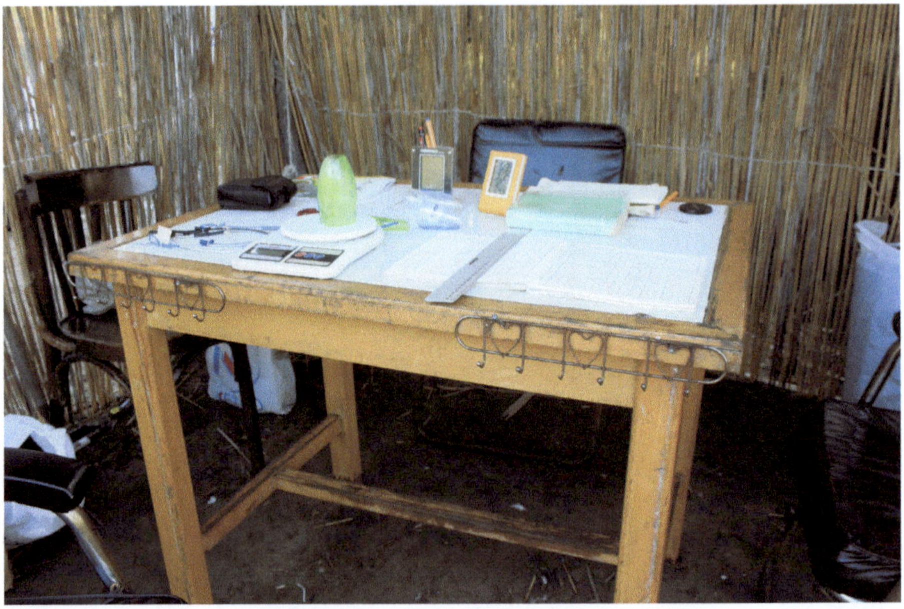

Figure 3.16-1: Seasonal bird ringing stand – Ashtoum, Egypt. Photo W. Kania.

Figure 3.16-2: Seasonal bird ringing stand – Burullus, Egypt. Photo L. Maksalon.

Figure 3.17-1: Provisory ringing stand – Barberspan, South Africa. Photo L. Pilacka.

Figure 3.17-2: Provisory ringing stand. Wadi Dana spring, Jordan. Photo I. Rząd.

Figure 3.18-1: Temporary, but well equipped ringing stand in a dry area. Natural shadow. Wadi Dana autumn, Jordan. Photo P. Busse.

Figure 3.18-2: Natural pool attracting the birds. Jordan. Photo K. Alomari.

4 Arrangement of the Netting Area

4.1 Land Habitats

Nets should be located in places with good catching prospects along a control path of reasonable length. Some practice is needed in optimizing a new location. When beginning work on new catching-grounds, even experienced ringers may not initially succeed in this task. Do not hesitate to adjust your location when you become familiar with local bird movements. Here are some general rules, which might be helpful:

1. Decide whether the catching area coincides with a path of active diurnal movement of birds, or, alternatively, if it is an ecological island, a real island in the sea, a big lake, or isolated biotope surrounded by habitats unsuitable for a group of birds of interest (Figure 4.1). Diurnal migrants frequently fly along guiding lines like seas, riverbanks or shores, strips of bushes between forests, and so on. The same applies to many nocturnal migrants, especially during peak days, but in a less visible way; they move from one bush or tree to another.
2. Nets placed within an area of active movement must be oriented perpendicular to the main direction of movement. Nets placed at other angles will not catch optimally. At times, the stream of migrants is so narrow that the only sensible thing is to concentrate nets within this corridor. You can even locate nets one after another at an interval of around ten meters (birds can easily to be caught into the next net when flying down after missing the first one). Such peculiarities of the area will be detected during peak days of migration only. In transient locations, it is advisable to locate the nets in front of more conspicuous bushes or just behind them (or both).
3. At island-type localities, most attention should be paid to differentiation of micro-scale habitats. Locate nets between two bushes, across paths of trees and bushes, at borders between different habitats etc. (Figure 4.2). Look for berries and seeds attractive to different species as well as water pools (Figure 3.18-2) which are attractive for most birds, especially within a dry area or during a period of drought. When the area is exposed to wind, look for localities that are not exposed to these prevailing winds; the best solution is to position nets so that some nets will be always protected from the wind, irrespective of wind direction.
4. Selection of the catching locality in the Middle East and Africa countries has some specific constraints. Regarding ecological type, the site usually has an island-type character – this includes groups of bushes, commonly *Tamarix* sp., *Nitraria* sp., patches of reeds near small local water bodies, sewage farms or cultivated orchards, and olive-tree plantations. So, the general philosophy of nets distribution is as it is for island-type localities. The difference depends, however, on general geographical location – in some places we can expect only a short-

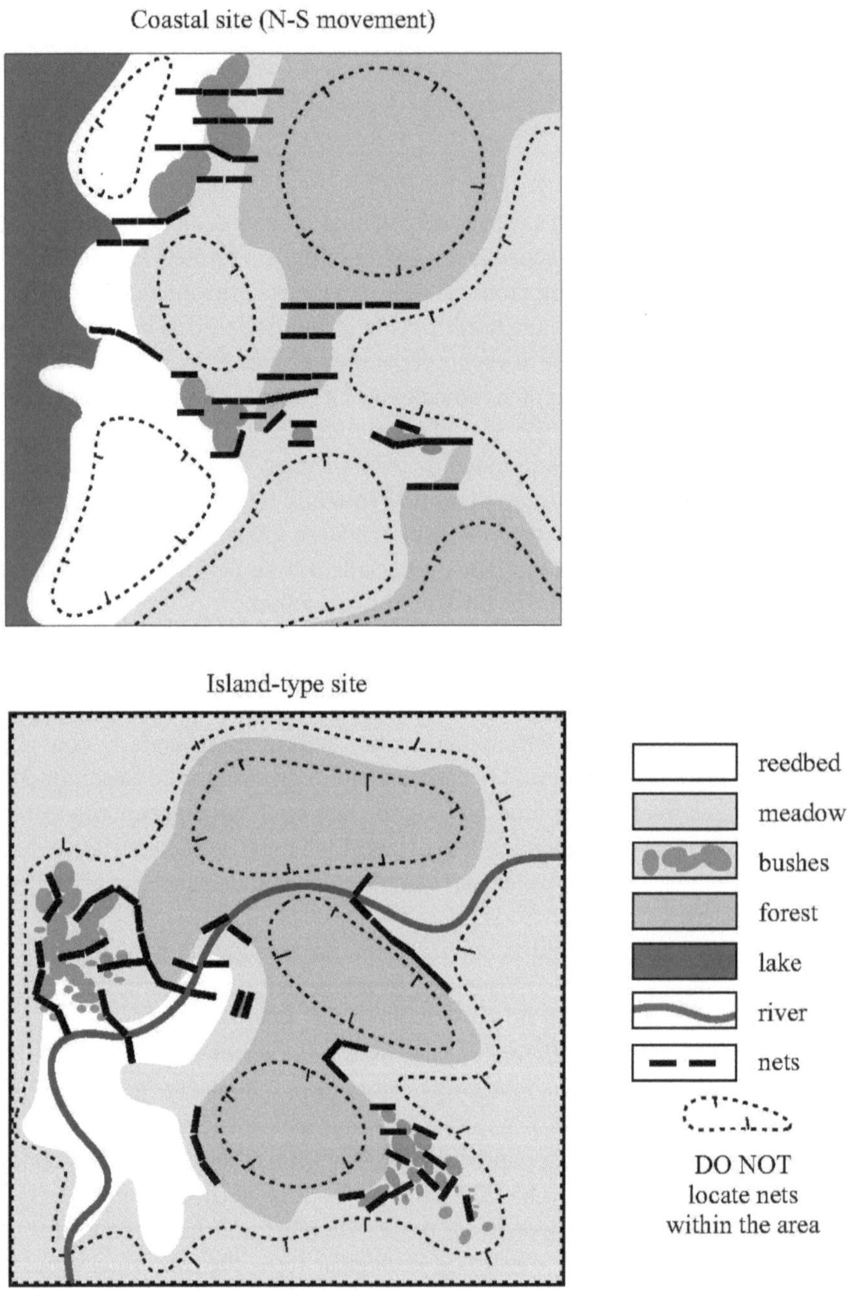

Figure 4.1: Example of properly distributed nets in two types of sites: "coastal type" site where the coast is a guiding line for migration and "island type" site with no directional movement of birds.

Figure 4.2-1: Typical situation of double net in front of bushes. Manyas, Turkey. Photo P. Busse.

Figure 4.2-2: Location of catching area in an island of bushes at a steppe. Olenevka, Ukraine. Photo L. Maksalon.

stop behaviour of nocturnal migrants; they land at the end of a night, stay a day and depart next evening. In such places, we can expect only one to two hours of good morning catches; the birds move around in very short time and then rest without changing place, and distribution of nets then plays less important role; there is no sense to look for feeding places. At a place where real stop-over can occur, like when birds are looking for good feeding conditions (e.g. vineyards, fig or olive plantations, and sites around drinking water available), one may find good catching prospects. Other constraints have a logistical character: living conditions, local transport, sun shelter, etc.

Nets may be used in singles or connected in rows. As a rule, single nets are more effective than the same number of nets built into rows. However, local configurations of vegetation may necessitate the use of rows instead of single nets. Rows are frequently used in more monotonous habitats, like reed beds, young tree plantations, etc. Where nets are set in long rows, a zigzag configuration is better than a straight line. Remember that nets should be easily accessible, for both birds and for ringers, from both sides, and the net should be located far enough from nearby vegetation. This allows the bird to achieve the flight speed necessary to "open" the shelf of the net, and the ringer will be able to remove the bird without entangling himself and the net in e.g. thorny bushes. When nets get entangled in e.g. alder cones or dog rose twigs, the mesh will inevitably take damage.

The length of the control path depends on the locality, number of nets, and anticipated number of staff available during peak days. Longer paths allow for a better selection of efficient netting locations, but when there are a lot of birds, the control will be too long. It is a good idea to make two or three shorter control paths controlled by different people or by one person (one after another), but always in the same order. This minimizes the potential number of caught birds after the control path, so he/she can come back to the laboratory, leave the birds, and then continue to the next path. In general, a control tour should not last more than fifteen minutes after few birds are caught. Control paths going through reed beds and marshes should be much shorter than paths in dry habitats.

Establishment of a new catching stand requires a few steps:

1. Walk around the area and carefully choose the best localities for nets. Try to look at the environment from the point of view of a migrating bird: which strip of bushes leads in the right direction for continued migration? Where would a bird hide when a raptor approaches? Where is their food and water? Consider the number of nets at hand and the expected length of the control path – should this site have one path or several?
2. Try to connect the nets selected by a path to be as short and straight as possible. Avoid steep hill slopes. Walking up and down hillsides may be harmful to the birds and to the ringer when the ground is wet, for if there is large number of birds, one may have to run along the path. If possible, avoid crossing ditches and

places easily flooded by rainwater. Furthermore, the full surface of any net or any net row should, if possible, be visible from the control path (Figure 4.3). If this requirement is complied with, it will only be necessary to approach nets when birds are actually caught and during evening and night controls. The path may run along the nets or pass their ends, i.e. avoid crossing them perpendicularly. The continuous need to pass under the bottom net string may damage the net, and the ringer could lose time when in a hurry.

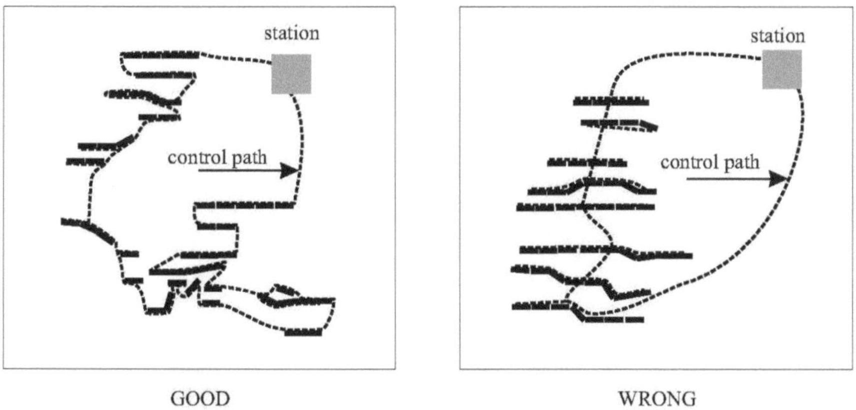

Figure 4.3: Correctly (left) and incorrectly (right) arranged control paths.

3. Thoroughly clear up the selected sites, prepare and erect the net poles in proper places; you will save time and not have to clear nets entangled in twigs, etc. when erecting them. Also perform a preliminary clearing of the control path.
4. Erect the station laboratory stand. When the first nets have been established, you will usually have good catches of local birds; readiness for this at the station laboratory is necessary, so erect it first.
5. Erect the nets and start catching.
6. Carefully monitor the catching efficiency of nets and the behaviour of birds around them. In many cases, it will be prudent to adjust the location of nets and to correct the course of the control path. This is rule rather than exception: the human eye is not a bird's eye.
7. When net locations and the control path are thoroughly fixed, clear up both nets stands (Figure 4.4) and the whole course of the control path. Consecutive adjustments may become necessary later on when e.g. new species start to migrate. When a net is entangled in a twig or any ground plant, remove the obstacle with a clean cut (land-owners, farmers, hunters: note the traces of your activities!). Particularly, the ground under the nets must be well cleaned – birds caught in the bottom shelf should not be entangled or hidden in ground

vegetation. This may cause their death when the weather is cold and wet, or they may be overlooked in darkness. Furthermore, for your own convenience and to save time: clear the control path. You should be able to walk along it with ease in order to avoid branches, twigs, fallen trees etc. When the ringer is in a hurry, an eye may get hurt by a twig, a leg broken over a branch, and the birds may even get killed when a bag hits the ground. And do not laugh here - this is the truth!

Starting a new season in a familiar area is much easier – you only have to clean young twigs and plants that sprouted in previously cleaned spaces.

Figure 4.4: Correctly (left) and incorrectly (right) arranged net stand. Arrow points the bird that may be overlooked.

4.2 Wetland Habitats

When arranging the catching area in wetland habitats, the procedure differs between pure reed beds and wetland with mixed sedge-reed-willow vegetation. A good catching opportunity is one objective, and reasonable time for control walks is another. The latter depends mainly on the quality of the ground. On permanently moist terrain, paths made of the boards fixed to wooden poles may be necessary (Figure 4.5). Where there is no possibility to use permanent paths, the speed of control walks in swampy ground covered by water will be low and the length of the path must be limited. Note that even apparently firm ground will become difficult to pass after being used a couple of times, this is particularly important to know when the path goes through standing water. The effectiveness of catching within the reed-beds differs from place to place. As a rule, the most effective nets are the ones located at the borders between reed and open water or bushes and trees. This border zone is the preferred habitat of the Reed Warbler. Single willow bushes attract a lot of birds, and their sides are usually good catching places. Linearly distributed (e.g. along the ditches) small bushes and trees always provide good catches; put the nets across such lines. In the core regions of monotonous reed-beds, the catching result is markedly lower. Open places with dispersed, low reeds and sedge have very low catching efficiencies. When rows of nets are erected within reed-beds, they should not be arranged as a straight line.

When nets are erected in wet habitats, their stretching and maintenance need extra attention. "Anchoring" strings attached to the end poles of a row will guarantee permanent stretching. The net should be set so high that birds caught in the lowest shelf will not get submerged in water, and the security margins must allow even for heavier birds (e.g. the Water Rail). Keep in mind that wet nets are longer than dry ones.

Figure 4.5-1: Comfortably arranged nets line through muddy area. Die Reit, Germany. Photo R. Lille.

Figure 4.5-2: Setting nets for catching waders at open wet area. Volturno plain, (Caserta), Italy. Photo W. Meissner.

4.3 Documentation of the Netting Area

When the catching area has been arranged, basic documentation must be made. This includes mapping the area with a description of the habitats. All net locations should be indicated and numbered, independently of the custom applied: noting or not noting net numbers in a ringing documentation. A list of erected nets by numbers must be made with descriptions of the net parameters (if differentiated) and the timeframe(s) of operation. Any additional changes should be written in this document, e.g. date of removal or addition of nets. When changing the net location, give the next free number as the number of a new net, e.g. "2 Sept. net no. 4 moved to 21".

Apart from a general description of the catching area, which is obligatory routine, a detailed description of the net surroundings within 20 m on each side should be done if the intention is to make a more detailed study of habitat preferences for a special project. This should be done separately for spring and autumn seasons.

The net location habitat coding after idea of W. Peach (*Manual of Field Methods* – Bairlein, 1995), modified (optional):

1. **Habitat type** (1 letter code)
   R – habitats with reeds
   S – scrub
   W – woodland
   X – other

2. **Habitat elements** (2 letters code)
   For habitat types R, S, W habitat details are coded:
   P – reeds (*Phragmites* spp.)
   T – reedmace (*Typha* spp.)
   J – rushes (*Juncus* spp.)
   C – sedges (*Carex* spp.)
   B – bushes
   H – herbs
   G – grass
   L – broadleaf trees
   F – coniferous trees

   Uniform habitat is coded by doubling the basic code, e.g. PP – pure reeds; mixed habitat is coded as two-letter code giving information on two dominant elements, e.g. PT – mainly reeds but with reedmace, TP – mainly reedmace, but with reeds etc.

   For X coded habitat type separate two letter code:
   MT – mountaintop
   ND – heathland
   AB – acid bog
   FM – farmland
   SM – salt marsh
   XX – other special habitat (describe in comments)

3. **Height of vegetation** (1 number code)
   Code average height:
   0 – less than 1 m
   1 – 1-2 m
   2 – 2-3 m
   3 – 3-6 m
   6 – 6-9 m
   9 – more than 9 m

4. **Presence of water** (letter/number code)

No water
N0 – dry
N1 – wet
N9 – dried out (water was earlier in the season, but now dried out)

Standing water
S0 – depth less than 10 cm
S1 – depth 10-30 cm
S2 – depth 30-100 cm
S3 – depth more than 100 cm

Flowing water
F1 – small stream
F2 – river

5. **Fruit** (1 number + 2 letter code)

Presence of fruits
0 – no fruit
1 – some fruit
2 – much fruit

Type of fruit
Every fruit has two-letter code – list two commonest ones; when more than two – code MX; when unknown – code XX.
JU – Juniper (*Juniperus* spp.)
TA – Yew (*Taxus* spp.)
IL – Holly (*Ilex* spp.)
EU – Spindle (*Euonymus* spp.)
RH – Buckthorn (Rhamnus catharticus, Frangula alnus)
RU – Brambles (*Rubus* spp., includes raspberry, blackberry, strawberry)
RO – Roses (*Rosa* spp., includes dog rose and sweet briar)
PR – Cherries and plums (*Prunus* spp., includes blackthorn, wild cherry)
CR – Hawthorn (*Crategus* spp.)
SS – Sorbus shrubs (*Sorbus* spp., includes rowan, whitebeam)
RI – Gooseberry family (*Ribes* spp., includes blackcurrant and redcurrant)
HI – Sea Buckthorn (*Hippophae rhamnoides*)
VI – Mistletoe (*Viscum album*)
AS – Strawberry Tree (*Arbutus unedo* and *Rhododendron* shrubs)
AE – Ivy (*Aralieceae* family like *Hedera helix*)
OL – Olive family (includes *Ligustrum vulgare* and cultivated *Olea europaea*)
SM – Nightshades (*Solanum* spp., includes *S. nigrum* and *S. dulcamara*)

CA – Honeysuckle fam. (*Caprifoliaceae* incl. *Sambucus* spp., *Viburnum* spp., *Lonicera* spp.)
SA – *Salvadora* spp.
NI – *Nitraria* spp.

6. **Habitat management** (1 letter code)
O – no management
N – normal forestry/farming management
R – main vegetation completely cut back at least once per year (e.g. burning or reed cutting)
C – coppicing
G – grazing
M – artificially managed (e.g. for monitoring purposes)
X – not known

For easier reading groups of codes can be separated by dots, e.g. S.BH3S1.1RUSO.O

5 Using Catching Devices

5.1 Netting

Nets must be maintained during the rounds of net control. This includes checking the tension and keeping the nets free from leaves, twigs, bigger insects etc. (especially dung beetles, which can do serious damage to nets). When a continuous catching routine has been established, special attention should be paid to the net maintenance in the evening or when checking the nets for the last time. A thorough cleaning of the nets before the last check is the best way to prevent birds from being overlooked in hidden positions; a passerine will die if left in the nets overnight. Additionly, the nets will be at the peak of their catching ability the next morning when a new wave of night migrants arrives. Also, clean the nets after a storm or heavy rainfall. They may be full of leaves and twigs after such events.

If the nets are pulled down after the morning catches, they must be cleaned up before closing, nothing worse can happen to them, except for a cow or an elephant, than being closed with leaves and twigs and then erected anew in darkness before the following sunrise! The "alarm" closing of nets (see *Laboratory Working Routine* - p. 118) does not allow to clean nets in beforehand, so the next opening could well be very difficult, time consuming and damaging to the nets, especially to the thin ones. This is one of the reasons why it is not recommended to close nets on a peak day. The nets are closed by putting all net ears together, so that birds cannot get accidentally entangled. This can be done by using clips, as for linen, to fix the netting. Wrapping the net around strings, commonly used, is the worst solution possible: the net become like Indian "*bal-chatri*" (trap for raptors) and a bird sitting on the net could be entangled as in trap's loops, especially when already ringed. During a windy day, wrapped net is more and more twisted and then is very difficult to open. After closing the nets for the day, they should be opened in the afternoon or evening rather than in the morning. The reasonable solution is to open nets around two hours before sunset, then make one normal control of nets and then make "night" as when nets are open day and night around. These two controls frequently will give us quite good catches as the evening movements of birds in many localities, including "short-stop sites", are well pronounced. The other advantage of the solution is that it will make possible a more efficient catch of birds landing before sunrise without disturbance from ringers opening the nets and, least but not last, it will allow the ringer to sleep longer. When opening the nets in the morning, in darkness, at least one hour before the sunrise, the nets should be properly opened and stretched. This needs a lot of time and training.

5.2 Extracting Birds From the Net

Various removing techniques are in use. They are differentiated by effectiveness (speed of removal) and safety for the bird. The technique, in contrast to "natural" removal by a layman, i.e. "no rules" – trial and error method, depends mainly on the accepted standard for holding the bird. One common technique is based on holding the bird with the tarsal joints between the fingers of the right hand, or worse, of the left hand. This technique is not a quick one. Although it allows a qualified ringer to remove the bird safely, beginners frequently cause injuries to bird legs or remove the birds too slowly. The technique described below is quicker, much safer for birds, and in practice, less complicated than its description.

First of all, the standard holding position of the bird should be trained. It is shown in Figures 5.1-5.2. The point of departure is always right-handed handling. Right-handed handling was chosen as standard procedure, despite the left-handed minority (sorry!), since the right hand fingers are better capable of holding the birds safely, meaning that it does not escape, and safely, for the bird, when it is removed and handled in a hurry. Left-handed handling is allowed for a left-handed person when removing birds, but at the next step, during laboratory work, the right hand holding must be used (once more, sorry!) for compatibility reasons (see *Laboratory Working Routine* – p. 118). It must be noted that handling the bird for demonstration and making documental pictures is different and it is shown at Figure 5.3.

Figure 5.1: Standard holding position of a bird (description in the text).

Figure 5.2-1: Standard method of holding a bird during ringing and examination of feathers. Wadi Allaqi, Egypt. Photo I. Rząd.

Figure 5.2-2: Standard method of holding a bird during examination of feathers. Wadi Allaqi, Egypt. Photo I. Rząd.

Figure 5.3-1: Correct handling of birds for demonstration and photographs. Comparing two small birds. Kopań, Poland. Photo unknown.

Figure 5.3-2: Correct handling of a bird for demonstration and photographs. Burullus, Egypt. Photo P. Busse.

Bird removing routine (Figure 5.4):

1. Determine from which direction the bird has entered the net. If this is difficult, apply the rule that the bird is on the side of the net where it has its belly. Remaining on one side and try to remove a bird that is on the other side is, although still possible, not convenient and not recommended for beginners.
2. Open the net pocket using left hand and hold the bird's body as deep as possible with right hand fingers.
3. Take out the bird with the net pocket toward your body.
4. Make sure that the net threads do not lie in a leg groin - remove them if so, otherwise it will be impossible to remove the close-lying wing.
5. Resolutely, but gently and carefully, pull threads going from the bird, using first three fingers of the left hand. Threads should be pulled one by one at a distance of at least 5 cm from the body. In most cases, wings and head will be removed quickly, but sometimes you must remove the head separately. Be careful, as the head being entangled into crossed threads is the most dangerous for the bird. The threads may be pulled relatively strongly, but not in a sudden outburst; bird's wings are very movable in all directions when pulled quietly. However, they can be broken too. At this stage of removal, do not care about an entangled bird's legs!
6. Hold the bird with the standard grip (Figure 5.1), as described above, using the right hand, the same as you have used for holding the bird's body previously - a comment important to left-hand removers! Changing the hand holding the bird is one of the biggest possible mistakes when removing it; in most cases, you will entangle it more than it was previously.

Figure 5.4: Bird removal routine (description in the text).

7. Freeing the legs - Most birds held with the standard grip will try to escape and free their legs from the net by themselves. Now, you are ready to put it into a bag. If the bird is not that kind, take the leg using your first and second finger of the right hand and hold it at the tarsal joint, not above it - you may break the leg! With the first fingers of the left-hand try to make threads slip by pulling them cautiously along the tarsus.

Where net threads are not extremely thin or hard, around 90 percent of all individuals caught will be removed quickly and without problems in this way. Some birds, however, get entangled in more complicated ways and individually practised techniques must be applied. A problem we often meet occurs when a thread gets hooked up on a tongue spur (especially frequent when thrushes are caught). In such case you have to hold both legs of the bird, since the kicking of the bird may injure its tongue (the bird will often make the tongue bleed by its own force). When legs are fixed, pull the thread backwards and upwards, over the spur, this will often suffice to free the bill. A very thin twig or a straw can be helpful sometimes Figure 5.5).

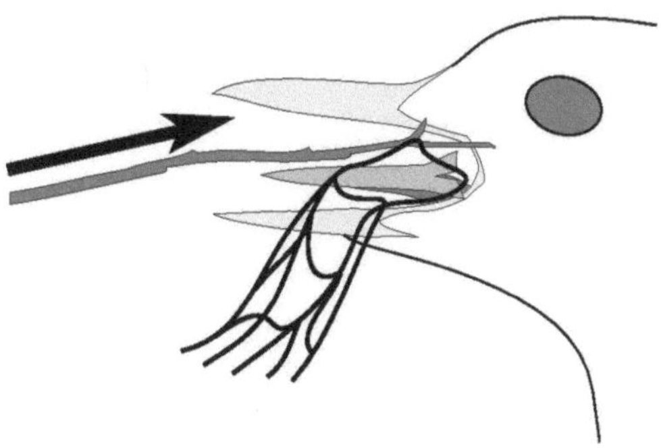

Figure 5.5: Removing a net thread from a bird's tongue.

At the end of the freeing procedure, you should always hold the bird in the standard fashion. Finally, put the bird into a bag, close the bag by pulling its string, slip down the lock (Figure 3.1) and hang it up on a special hanger on your neck (Figure 3.2) or on an eyepiece of a binocular. This last is a very practical custom: many ringers have a binocular ready to observe a rare bird, and it is a useful hanger for bags. When more birds of the same species are caught, closing the bag after each individual becomes impractical and time consuming. The solution is to hold the bag closed with the fourth and fifth fingers of the left hand (Figure 5.6) and remember that you use only three first fingers to the next bird removing (see point 5 above). When the bird is in your right hand, add it to the previous ones in the bag. When the standard number of individuals is in a bag (see p. 33), close the bag and use your hanger. For goldcrests you can, when you have a good training, use a special procedure - after removing the birds, you can collect them in the right hand, holding them by your fourth and fifth fingers, then remove another one and another one. With five of them in the hand, put them into the bag. When this has been repeated three times, you will have the standard number of goldcrests inside. The method is very quick and it seldom happens that a bird is clever enough to escape. However, the method is not good for long-tailed tits as they have "slippery" plumage and even third individual is not easy to keep.

Figure 5.6: Holding the bag during removing the bird when many birds are caught. Label with number is used when numbers of nets are noted according to the station routine.

5.2.1 Special Tips When Removing the Birds

Some birds, when caught, may hurt your fingers or even your eyes.

Hawfinches and shrikes may easily injure your hand by the strength of their bills, so the first thing when you remove such a bird, fix its head.

Raptors and owls hit mainly with their claws (Figure 5.7), and they are very quick. Surprisingly, their hook-like bills are usually not as dangerous, although there are some individual exceptions! The first thing to do when starting to remove a raptor or an owl is to hold them by the tarsal joint of both legs; this is an exception to the

Figure 5.7-1: Raptors head. Kopań, Poland. Photo W. Busse.

Figure 5.7-2: The leg of the individual above with a blood sample of the ringer. Photo W. Busse.

removal method! Owls look as if they are sleeping in the net, but the most dangerous is the first moment of removal - their legs are electrifying like lightening. If, despite your care, the bird catches some part of you by its claws, remain calm and slowly turn the leg along the tail to the bird's back (Figure 5.7) – its claws will automatically open, due to an anatomical peculiarity of the leg. The same may help when the claws are much entangled in the net and you are unable to remove threads.

Figure 5.8: Freeing up the leg when the bird with strong toes keeps the net. Use the same principle when the Jay, raptor or owl catches you.

Jays use both methods of fight, bill and claws, and your response must be to use both tactics mentioned above.

Tits are irritating by their pinching (Figure 5.9).

Special note: little bitterns are dangerous to your eyes when handled; they have surprisingly long necks and may hit your eye suddenly! The same applies to all herons and bitterns (Figure 5.9), and be careful with moorhens and coots as well!

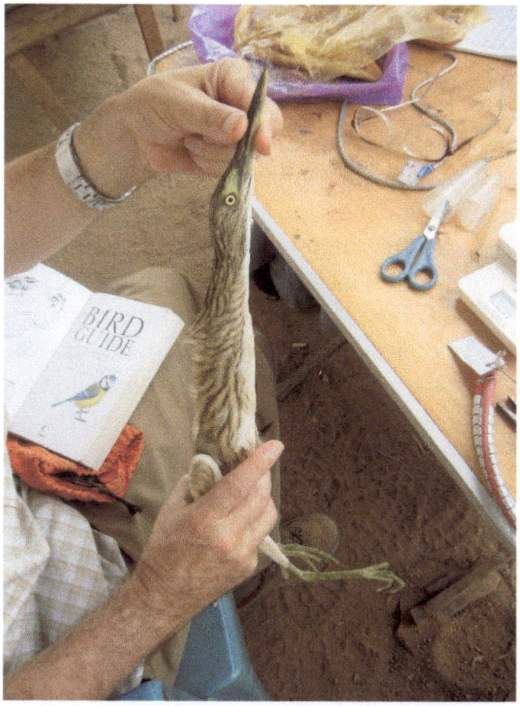

Figure 5.9-1: Neck length of the **small** heron – be careful. Wadi Allaqi, Egypt. Photo I. Rząd.

Figure 5.9-2: A tit pinching the ringer. Location and photo unknown.

5.3 Standard Set of Nets

The net controls should be done every hour at the beginning of every full clock hour, with start and end times depending on the latitude of the station. The first control should begin at (or up to) half an hour after sunrise. The last control of the day, after the listed ones, should be done at darkness, no later than 0.5-1.5 hour after the preceding one. In the Middle East, the first control is around 6.00 and the last 18.00-19.00. In these regions, a good time to start the night control is when bats start to move. This is a little bit before full darkness, but this comes in the region very quickly; if you start as advised above, you finish the control in full darkness. When owls are migrating and special owl nets are used, night controls performed at two-hour intervals are necessary. If no owls have been caught up till midnight, the next control may be skipped.

When the weather is wet and cold (or very hot, above 30 °C), the frequency of checks should be higher in order to avoid losing birds. The birds caught during additional checks should be treated as birds caught at the next regularly scheduled check.

The control walks should always be done in the same direction along the control path. This results in regular visits of the same nets and is of great help in alarm situations (see *Alarm Routine*, p. 123). If there are groups of nets regularly catching more than others along the path, they should be visited later on the control walk. This is important when many birds are caught; they will then, on average, be transported a shorter distance (this is for their, and your, convenience).

There are a few rules that should be followed when the nets are checked. They derive from the general handling system that gradually evolved when great amounts of birds were caught in the past. Applying these rules will help you in handling the birds and minimize losses. The advantages from adopting this handling routine are most obvious on peak days of migration.

1. Start each control walk with a sufficient number of bags. This is especially important at the first control walk in the morning, when unexpected rushes may sometimes occur. Returning to the station for a new set of bags will ruin the rhythm of controls. Remember this rule during all controls throughout a day, as birds sometimes may come as a big surprise (esp. tits and starlings). If, however, you must return for new bags, go directly to the net where you used your last bag; do not remove newly caught birds from the nets already controlled, otherwise you risk not to be able to finish the control walk in time.
2. Prior to the transport of birds, rigorously select them by species: allow only one species in one bag. Try to remember which species is in which bag – the best is to use colour bag codes for the most common species, e.g. orange ones for robins, blue for blue tits etc. When there is not enough colours, hang e.g. goldcrests on a left hook of the hanger while robins in a bag of the same colour hang on the right side. This rule is useful both during control (adding new birds to previously

caught) and, especially, at the laboratory where all persons coming from different control paths must hang the birds selected by species (see *Laboratory Working Routine*, p. 118). When many birds are caught in one net or rows of nets remove them by sides of the net and by species (do not close a bag before filling it – see above). It is much less time-consuming to remove all birds from one side of the net and then go to another side than to remove birds "on reverse". If two persons remove birds from one net simultaneously, one of them should specialize on one and the other one on another common species; you will have less strain to remember "who's who" in the bags.

In some programmes, a net number should be noted for ringed birds. There are two main reasons for such data collection: the wish (1) to study habitat preferences and (2) to follow territorial behaviour of the particular individuals. In the first case, you only need to know that e.g. four robins are from net no. 3 and two more from net no. 11, but not that exactly this individual is from net no. 3 and that one from net no. 11. So, you can put them all into one bag and note on a piece of paper "4 from no. 3, 2 from no. 11". In the second case, you have to keep robins from e.g. net no. 5 only in one bag; you have to carry much more bags along. One good advice is to have special plastic labels with net numbers. The label should have a hole of such size that the bag string can be easily laced through (Figure 5.5). The labels are stored on hooks at the net poles. After putting the bird caught in a particular net into a bag, you place the label on the bag string and you do not need to remember or write the number of the net. After a few control walks you will, however, be forced to redistribute labels collected at the laboratory to the proper nets. This idea can be used also for marking the bird species being transported in different bags.

3. Remember that there are limitations to the number of individuals allowed in one bag. It depends on bird species and circumstances. Standard numbers are listed in Table 2.1.

   The central, bold numbers are valid assuming that both bags and birds are dry, in good condition, and not expected to wait too long for ringing. When birds are wet or when they have to wait for ringing more than one to two hours, reduce the number (first numbers valid), while if you have just finished one control walk and birds will soon be shifted to the waiting boxes, you may fill the standard bag with higher number of individuals (last numbers valid).
4. Avoid transporting full bags or hanging them in a larger number onto one another. When you have too many full bags, hang them not only on the hanger or binocular but on forearms as well. If you have full bags at the start of the control walk, and you know that you will pass nearby at the end of the walk - hang the bags on a tree (in shade!) and do not forget to take them on your return way. In a case when you are forced to run with full bags, support them from the bottom to protect them against hitting one another and your body. So, be very careful when you transport larger numbers of birds simultaneously; your mistakes could cause

their death. Wrong handling of full bags is the most common cause of losses among birds caught on peak days.

5. Immediately after reaching the laboratory, hang bags selected by species on appropriate hooks (Figure 3.14) and/or, after an order of the chief ringer, shift some birds caught from bags to waiting boxes. Report to the chief ringer any problem expected on the next control, e.g. "a lot of tits are coming".

   Sometimes, many birds can be caught during last control (particularly in nets erected in reeds nearby roosting places of starlings, swallows or wagtails); so good lamps (including headlamps) should be available as well as a good source of light in the laboratory.

5.4 Special Netting

Within the framework of the "passerine station", some degree of special netting can be performed. The special nets for catching Passerines are sometimes built to catch the birds that make ringers nervous - by flying too high to get caught into the standard nets. There are different constructions. One type consists of a normal size net pulled up and down on strings along high poles (e.g. Figure 2.4-2), another of very high nets (eight to ten shelves) slipped down when birds are caught in higher shelves. Such special passerine nets could make ringers happy by catching a few more birds and/or some birds that are rarely caught in standard nets. However, they are very laborious in action, as the net must be lowered in order to remove birds caught above the range of the ringer's reach. Additionally, when the net has a lot of shelves, they will close when pulled down, and the bird usually gets much entangled. Use of such nets, as an addition to the normal set of nets, is up to the individual ringer. During the period of potential mass catchings, such nets should not be in use, or at least they should be pulled down when a rush of birds is anticipated. Otherwise, they could be a cause for substantial problems with bird safety (see – *Laboratory Working Routine* – p. 118). At times, normal-sized (usually doubled-length) nets made of 25 mm mesh netting (which are more efficient in Thrush, Sparrowhawk and Cuckoo catching) may be used as an addition; they provide good "protection" against raptors at migration sites. Such nets (although they are not on par with normal nets for monitoring data) need not be closed during the peak days.

Special nets intended to catch larger birds, usually raptors and owls, are sometimes used at the "passerine stations". These are nets made of big size netting (40-80 mm mesh) and much higher than the standard ones. Because of the big mesh, such nets usually do not catch small birds, but when they are caught, they may get extremely entangled! These are the source of peaks in catching, but if the rush of small birds is connected with the Jay migration, they can cause problems. If such nets are used to catch owls during their migration, they are very efficient and should be controlled during the night at two-hour intervals during good weather. If there are no owls caught till midnight, subsequent controls could be skipped as the next catches could be expected end at sunrise.

5.5 Attracting the Birds to Nets and Traps

If the number of birds caught at a defined site is unsatisfactory to the ringer, one may start to think about how to attract more birds. One possibility is that we must ring more. For scientists, ringing birds for research purposes adds less psychological pressure, but still, statistical treatment of the data requires rich data files. From this point, we must decide what our goal is really, as the "*more, more and more...*" idea has its good and bad sides. If the main goal of catching is producing only ringing recoveries or collecting biometric data, the idea is acceptable, but still remember that ringing outside of breeding or wintering seasons without collecting reliable migration dynamics data is of relatively low value for ringing recovery evaluation. Collection of sufficient data sets for special studies can be in contradiction with mass ringing; if all staff must be engaged in removing the birds from the nets and ring them, collection of special data, e.g. performing orientation tests, collecting blood samples etc., becomes impossible. As a result, too many birds are caught, and we have serious gaps in the special study data, while peak days data could be of major value to these studies. In such a case, we need to reduce number of the birds caught in favour of having good samples for special studies. Reducing catches in rush days only ruins the migration dynamics data. So, frequently, we have serious dilemmas without optimal solutions, and how to arrange the work at the station in such a way that different needs could be met fail to be realized. Despite of some problems caused by catches of too many birds (we frequently want to catch more birds), always think in advance about negative consequences that can occur in peak days. There are a few solutions:

Use more nets. Note, however, that this solution needs more nets, more net-poles, more work if closing the nets is in the standard routine of work and, usually, setting longer control paths.

Use thinner nets as a standard. Note, however, that this causes much harder work with removing birds, thus there is a need to decide optimal course for birds safety (see comments in the *Alarm Routine*, p. 123).

Attracting birds to nets or traps by baiting them with food or water. This can be used if attracting will be continuous; this does not change seasonal dynamics of the species catching; useful in work using "active" Heligoland traps during migration time. This is a basic element for winter catching close to different types of bird feeders. Depending on the food served, different birds will be attracted, e.g. animal fat and oil rich seeds are very attractive for tits, nuthatchers, woodpeckers, as well for many other species, different seeds for finches and oat for Yellowhammer. It is less known that feeding birds during breeding time is very effective for ringing as parents use the bait for feeding chicks and guide freshly fledged ones to the feeder. By the way, this custom can save a lot of fledglings from dying when longer period of very bad weather occurs.

Attracting birds by setting decoys of owls near the nets (Figure 5.10-1). They are attacked by passerines and raptors.

Attracting migrants using so called "tape-luring" (Figure 5.10-2). This traditional name is derived from using in the past analogue voice registering and playing devices, tape recorders, for attracting the birds to nets or traps. Nowadays, there are digital recorders and mp3 players, running CD or memory sticks, in use. This is a very controversial method of attracting migratory birds to a certain place or a catching area by broadcasting bird voices using loudspeakers. There are three different, but frequently combined, procedures using this method: (1) while migrants are still on air (before a dawn), (2) during a day for attracting birds being within the area to a certain nets and (3) in the evening attracting birds roosting nearby to places where nets are set. For waders, there is used tape-luring at night (see p. 168). Migrating passerines, attracted by broadcast voices land, instantly in a vicinity of loudspeakers set among the nets and then moving around are caught even after the tape-luring is stopped. This is, however, frequently combined with the second procedure, and loudspeakers work continuously. A third procedure is used separately. A similar method is used during breeding time to catch territorial birds or to stimulate vocal responses of birds that are difficult to find during faunistic studies (e.g. owls). The technique of the tape-luring can base on car tuners supported by enough powerful loudspeakers or even small mp3 devices. According to technical solution and procedure, used batteries must have appropriate power capacity. The first procedure needs the most powerful loudspeakers and batteries at a distance from which the voice must be heard, and should be as long as 1-2 km. For attracting local birds and roosting ones, the demand for power is much lower. A CD contents, or nowadays, a stick memory file, must be set for loop running. Their contents depend on the procedure used and the goal of the tape-luring. Generally, more diversified voices can attract more species. It must be stressed that attracting birds by voices, frequently songs, has limited species specification; voices of one certain species attract not only this species, but, sometimes several others, even those that do not live side by side within the same type of habitat. This statement is very important when considering the birds' safety, and this is a source of discussions and controversies. In some countries, tape-luring is even prohibited or limited. The problem arises (mainly in using of the first procedure) when luring target birds while air-bone, or migrating. Migrating birds endogenously determine their orientation, all while maintaining timed changes in certain physiological processes. The time at which birds fly (between departure and arrival) is set by physiological abilities of an individual, and crossing migration barriers can be done only in certain weather conditions. The decision to stop or stopover depends on many natural cues to which the bird is evolutionary adapted to detect and respond. In such situations, a strong influence, such as that of an artificial stimulus by tape-luring, can change the status of the migrant and it breaks migration in improper, or at least in a non-

Figure 5.10-1: Owl plastic decoys for attracting passerines. Tanzania. Photo P. Busse.

Figure 5.10-2: Tape-luring equipment used for attracting swallows. Aras, Eastern Turkey. Photo P. Busse.

optimal, place and time. Landing locality and biotope depend not on natural bird preferences but on the place where ringer tape-lures. This can be strange to a species; reed bed birds can land in dry bushy area or even in open farmland (there are such known ringing tape-lurings to attract birds to beans fields!). In contrast, forest birds are forced to land within huge reed-beds. Such tape-luring is clearly not moral. It is clear that such a landing cause disrupts these birds, exposing them to raptors, lack of proper food and rest places etc. The problems depend on local situations; tape-luring of reed-bed birds into reed-bed or warblers to a mixed bushy and tree site cannot be perceived to cause problems to them, but remember that sometimes, low species-specific composition of voices are broadcasted by you. Another constraint that you must have in mind is the localization of your site in relation to migratory barriers that birds must cross: sea or desert, and to a lesser extent, mountains. On the borders of these barriers, on both sides of them, before and after, tape-luring with the first listed procedure should be totally prohibited; these localities are crucial for migrants. Tape-luring as attracting birds to a certain net stand during a day is less effective and, from other side, not so disturbing to migrants. Using tape-luring to move roosting birds to nets stands should be done with care; do not force them to places unsuitable for a safe roosting.

5.6 How to Arrange Trapping with Heligoland Traps

Special attention must be paid to the maintenance of Heligoland traps in order to make them really efficient. All ropes and strings should be rigorously stretched and the netting free from holes. Holes are dangerous to birds that may get caught and die. Other holes, even small ones, may serve as an escape to a lot of birds, especially when holes, sometimes one single broken mesh, are situated at corners of the terminal room. Birds, especially tits, observe individuals that escape and instantly follow in their path. Removal of birds from the Heligoland trap differs from freeing of birds caught in nets and depends on the construction of the trap. Birds may be removed by hand from a permanent collecting box, caught in the terminal box with a hand-net similar to that used for butterfly catching, or taken with a collecting box that is replaced with a new, empty one. Transport of birds collected at a Heligoland trap can be done in transport boxes, since the trap is the only place where the birds are collected (zigzag trap is an exception as there are a couple of collecting boxes). When many birds are brought together in one collecting box, "sorting by species" is necessary before the box is given to the ringing stand. Note that the potential catch of any Heligoland trap is huge, and all the rules of "quick handling" must be strictly observed. In Heligoland traps, where the birds are caught in collecting rooms, "hunting" for the last trapped individuals jumping from one wall to another is very laborious, but this must be done for the safety of these unlucky individuals. Only when a permanent, intensive flow of birds occurs can some individuals be left behind in the catching room. After the rush, all remaining individuals must be removed. A similar situation may occur in Heligoland traps with collecting boxes when some individuals hesitate to enter the box. The rules of their treatment are the same as described above.

The Rybatchy-type trap should be installed and erected in the following way. The netting piece is placed on the ground inside of the carcass construction. The first rope is fixed to the top of one of the first pair of pillars. It then passes through the rings on the trap and through the small block fixed on the carcass rope (near the top of other pillar) of the first pair and then it is directed toward the winch. The same sequence is repeated on the second pair of pillars. The ropes should be stretched as far as they will allow. The special marks on the stretched wires should be made at a distance of 0.5m from the top of pillars. Subsequently, it is necessary to release the ropes a whole turn, and the trap should be fixed to the ropes at these marks. After that, the ropes once again should be stretched as far as they will allow, and the trap at last will come into working position. To the third and fourth pairs of pillars, the trap is fixed by hand. The bottom of the trap must be fixed to the ground.

The entrance part of the trap must be closed down when a strong wind (more than 5° Beaufort scale) occurs. One person is able to close the trap within 15-20 minutes. Opening it again will take 30-40 minutes and more than one person should do that. Experience shows that the trapping efficiency is affected mainly by the wind direction. Contrary and side-contrary winds of moderate force reduce the height of the birds' flight; it is under such circumstances that the majority of migrating flocks are trapped.

The maintenance work at the Rybatchy-type trap requires at least three people. Cases where birds are injured or killed in this trap are relatively rare. They occur during very intensive migration when thousands of birds are trapped. The primary cause of death is overcrowding of birds in the collecting and terminal part of the trap. Ornithologists in general agree that the trapping with the Rybatchy-type trap is safe for the physical conditions of the birds.

For successful operation of the Rybatchy-type trap, it is necessary to choose a proper site for its construction. Careful observations of migrating birds under different weather conditions are very helpful, since the local habits of migrants may differ greatly. For example, on the Courland Spit genera like swallows, pigeons, crows (except jays) are trapped rarely, although migration of these birds is very intensive, whereas in Kazakhstan, at Chokpak, these birds are the most numerously caught in traps. Owing to its large size, the Rybatchy-type trap is not perceived as a place that birds must avoid. Therefore, no camouflage, e.g. by the special colouring of the net, etc., is necessary. During summer, some birds (especially chaffinches) are not only trapped repeatedly (up to several scores of times), sometimes they even build nests inside the trap.

Usually, maintenance of the zigzag trap is simple, and the trap should be checked every hour. However, during an intensive migration, a large number of birds may get into the chamber simultaneously, so the trap should be controlled more frequently.

When the birds in the chambers or baskets are of different sizes, it is recommended first to remove the bigger birds. Sometimes, one can find birds entrapped between the "wings" and the sides ("walls") of the trap. Such birds may easily be driven into the chamber or basket. When birds have been removed, one has to make sure that the "sleeves" are tied up again to prevent other birds from escaping from the basket or chamber.

6 Passerine Station Laboratory Methods

6.1 Species Determination and Coding

Species determination is undoubtedly a fundamental starting-point for ringing. For ringing purposes, a good key or "the bird in the hand" type of manual should be used. Common "guides to...", coloured books for field identification of birds are based on characters visible from a distance, frequently including behaviour and bird voices. These sometimes could be insufficient for correct determination of an individual catch; some field characters are no longer visible at the bird with disturbed plumage, and the voice of the bird crying when removed from the net have no similarities to natural calls. On the other hand, "the bird in the hand" manuals present species-specific characteristics, which are completely invisible in the field, e.g. details of wing-formula or colour patterns of single feathers. These details should be carefully noted when rare species or species difficult to determine are ringed.

There are two main types of determination procedures in use: a key system where alternative characters are listed in hierarchical order, usually a dichotomous form and guide system, where alternatives of different characters are given as sometimes long text describing more or less important details. The first, traditional system is easier to handle for less experienced workers that are guided to final decisions by the construction of the key. However, misunderstanding one step in the key sequence (or misunderstanding the description given) could lead to wrong species determination. This key determination must be confirmed by a careful study of the species description given in the guide form. Any doubts should be clarified in the early stages of discrimination. Guide systems are good for experienced ringers who are well acquainted with the guides' manner of description and for ringers who know which characteristics are the most important ones in the family of the individual caught. There is less confusion, and if there is any contradiction between characters, it is easier to assess the relative value of the characters in question.

Correctly identified bird species must be correctly noted in the ringing form. Because of the inconveniences of plain text writing, for ringers as well as for the person who loads collected data into data-file, different forms of species name codes are in use. They render the species name in a short form that is easy to write down under field conditions and is not time-consuming when typed on a keyboard.

For mnemotechnical reasons, any number code must be excluded from the field use; letter codes are easier to remember and less vulnerable to errors. The most universal in the international network is a five or six-letter codes based on scientific names and specially prepared to be error-proof. The main idea is construction of the code in two segments – **two** (in 5-letter code) opening letters from the scientific genus name or **three** (in 6-letter code) e.g. PA or PAR (from *Parus*) and three

FIRST letters from the species name, e.g. MAJ (from *major*) – that gives PAMAJ (5-letter code) or PAR.MAJ (6-letter code) for ***Parus major***. However, this simple, standard rule can sometimes lead to identical codes given to different species, e.g. **PHY***lloscopus* **TRO***chilus* and **PHY***lloscopus* **TRO***chiloides* or **ACR***ocephalus* **PAL***udicola* and **ACR***ocephalus* **PAL***ustris*, so in these cases, other procedures must be used. The additional, B-procedure uses three opening letters from the generic name and three **LAST** letters from the species name, thus the above-mentioned *Phylloscopus trochilus* is coded as **PHYLUS**, while *Phylloscopus trochiloides* as **PHYDES**. Note that code **PHYTRO** is not used anymore, since it is meaningless. Additionally, this procedure should be applied for coding the species that could have standard codes very similar when hearing by a person noting the ringing data into the ringing form, e.g. 6-letter code for *Calidris alba* using standard procedure would be CAL.**ALB**, while code for *Calidris alpine*, CAL.**ALP**, sounds very similar and erroneous codes could be easily written. To avoid typing errors when inputing data from a keyboard some codes differentiated only by one letter – being close to other one at the keyboard, e.g. TRI.**F**LA and TRI.**G**LA as **F** and **G** are side by side on the keyboard, are differentiated using the B-procedure, described above. Even the B-procedure does not solve all problems; some species codes that are created by the A-procedure are identical or very similar to other codes. To solve the last of the remaining problems after trial of B-procedure must be solved by C-procedure: using as the second part of the code three letters from the species name, but neither the first nor last ones, e.g. CARduelis flaVIRostris. Among all bird species listed in the EURING list of species, supplemented by species that can be normally found in the Middle East, there are 555 standard codes, 51 B-procedure codes and 8 C-procedure codes. The special procedure codes, as well as group codes, for visual observations are listed in Tables 6.1-6.2. If letter codes are used outside of the Western Palaearctic and the Middle East, local code sets must be elaborated and checked for possible doubling of codes and possibility of errors. Note that during last years, a number of genus names have been changed and the process is still causing troubles to many field ornithologists who are not familiar with current taxonomical developments. It must be taken into consideration whether introduction of the code sourced in the new name does not double the existing one and decide to create new code after B- or C-procedure instead. The same problem is with new species names when the well-known species is split into two or more items. E.g., separation two subspecies of the Stonechat (coded as SAX.TOR) – *Saxicola torquata rubicola* and *S. t. mauri* – into full species *S. rubicola* and *S. mauri,* that could be coded according to A-procedure as SAX.RUB and SAX.MAU accordingly, made the first code meaningless because of already existing code SAX.RUB for the Whinchat, *Saxicola rubetra*. This changed coding of Whinchat and European Stonechat to SAX.TRA and SAX.OLA, according to B-procedure.

Summarising, the species 5- and 6-letter codes **are created in the following standard way:**
Genus code: FIRST TWO (5-letter code) or FIRST THREE (6-letter code) letters of the genus scientific name.
The basic procedure A: *genus code* + FIRST three *letters of the scientific species name,*
The procedure B: *genus code* + LAST three *letters of the scientific species name,*
The procedure C: *genus code* + three OTHER *than above letters of the scientific species name.*

Non-standard code items, constructed according to special procedures: B and C, are listed in Table 6.1 and the group observation codes, for the birds not identified to the species level, listed in Table 6.2.

Table 6.1: List of non-standard code items according to:
B-procedure (genus code + three LAST letters of species name)

	5-letter code	6-letter code
Acrocephalus paludicola	ACOLA	ACR.OLA
Acrocephalus palustris	ACRIS	ACR.RIS
Calandrella rufescens	CAENS	CAL.ENS
Calidris alba	CAALB	CAL.LBA
Calidris alpina	CAALP	CAL.INA
Calidris canutus	CATUS	CAL.TUS
Calidris fuscicollis	CALIS	CAL.LIS
Calidris minuta	CAUTA	CAL.UTA
Calidris minutilla	CALLA	CAL.LLA
Caprimulgus ruficollis	CALIS	CAP.LIS
Carduelis cannabina	CACAN	CAR.INA
Carduelis carduelis	CACAR	CAR.LIS
Carduelis flammea	CAMEA	CAR.MEA
Chettusia leucura	CHURA	CHE.URA
Chlidonias leucopterus	CHRUS	CHL.RUS
Corvus c. cornix	CONIX	COR.NIX
Corvus corax	CORAX	COR.RAX
Corvus corone	COONE	COR.ONE
Cyanistes cyanus	CYNUS	CYA.NUS
Cyanopica cyana	CYANA	CYA.ANA
Emberiza pusilla	EMLLA	EMB.LLA
Emberiza rustica	EMICA	EMB.ICA
Ficedula narcissina	FIINA	FIC.INA

continued Table 6.1: List of non-standard code items according to:

	5-letter code	6-letter code
Ficedula parva	FIRVA	FIC.RVA
Lanius minor	LANOR	LAN.NOR
Locustella luscinioides	LODES	LOC.DES
Luscinia luscinia	LUNIA	LUS.NIA
Milvus migrans	MIANS	MIL.ANS
Milvus milvus	MIVUS	MIL.VUS
Oenanthe leucopyga	OEYGA	OEN.YGA
Oenanthe leucura	OEURA	OEN.URA
Phasianus colchicus	PHCUS	PHA.CUS
Phylloscopus trochiloides	PHDES	PHY.DES
Phylloscopus trochilus	PHLUS	PHY.LUS
Podiceps cristatus	POTUS	POD.TUS
Podiceps grisegena	POENA	POD.ENA
Porphyrio porphyrio	PORIO	POR.RIO
Porzana porzana	POANA	POR.ANA
Pyrrhocorax pyrrhocorax	PYRAX	PYR.RAX
Pyrrhula pyrrhula	PYULA	PYR.ULA
Saxicola rubetra	SATRA	SAX.TRA
Saxicola rubicola	SAOLA	SAX.OLA
Stercorarius parasiticus	STCUS	STE.CUS
Sterna paradisaea	STAEA	STE.AEA
Sylvia conspicillata	SYATA	SYL.ATA
Sylvia melanocephala	SYALA	SYL.ALA
Sylvia melanothorax	SYRAX	SYL.RAX
Tetrao tetrix	TERIX	TET.RIX
Tetrax tetrax	TERAX	TET.RAX
Tringa flavipes	TRPES	TRI.PES
Tringa glareola	TROLA	TRI.OLA

C-procedure (genus code + three not FIRST nor LAST letters of species name)

	5-letter code	6-letter code
Carduelis flavirostris	CAVIR	CAR.VIR
Ficedula albicollis x hypoleuca	FIAXH	FIC.AXH
Larus minutus	LAUTU	LAR.UTU
Parus montanus	PAUMO	PAR.UMO
Passer montanus	PASMO	PAS.SMO
Phylloscopus collybita	PHBIT	PHY.BIT
Sylvia communis	SYUNI	SYL.UNI

continued **Table 6.1:** List of non-standard code items according to:
OLD-codes **(codes used in the past)**

	5-letter code	6-letter code
Acrocephalus paludicola	ACPALA	
Acrocephalus palustris	ACPALS	ACR.UST
Acroceplalus palustris	ACUST	ACR.UST
Acrocephalus schoenobaenus		ACR.ENO
Acrocephalus scirpaceus		ACR.IRP
Calidris minutilla	CAMINL	CAL.LLA
Carduelis flammea	CAFLAM	
Carduelis flavirostris	CAFLAV	
Emberiza pusilla	EMPUS	
Emberiza rustica	EMRUS	
Ficedula narcissina	FINAR	FIC.NAR
Ficedula parva	FIPAR	FIC.PAR
Larus minutus	LANUT	LAR.NUT
Locustella luscinioides	LOLUS	LOC.LUS
Luscinia luscinia	LULUS	LUS.LUS
Mergus merganser	MEMER	MER.GAN
Mergus serrator	MESER	MER.RAT
Milvus migrans	MIMIG	
Milvus milvus	MIMIL	
Phasianus colchicus	PHCOLS	PHA.COL
Phylloscopus collybita	PHCOL	PHY.COL
Phylloscopus trochiloides	PHTROO	
Phylloscopus trochilus	PHTROU	
Podiceps cristatus	POCRI	POD.CRI
Podiceps grisegena	POGRI	POD.GRI
Pyrrhocorax pyrrhocorax	PYPYRX	
Pyrrhula pyrrhula	PYPYR	
Saxicola rubetra	SARUB	SAX.RUB
Saxicola torquata	SATOR	SAX.TOR
Stercorarius parasiticus	STPARS	STE.SIT
Sterna paradisaea	STPARA	STE.AEA
Sylvia communis	SYCOM	SYL.COM
Sylvia conspicillata	SYCON	SYL.CON
Tetrao tetrix	TETETI	
Tetrax tetrax	TETETA	
Tringa flavipes	TRFLA	TRI.FLA
Tringa glareola	TRGLA	TRI.GLA

Table 6.2: Group observation codes, used when the birds are not fully identified

Group	Code
Accipiter sp.	ACC
Anser sp.	ANS
Anthus sp.	ANT
Branta sp.	BRA
Buteo sp.	BUT
Circus sp.	CIR
Columba sp.	COL
Corvus sp. = *Corvidae*	COR
Falco sp.	FAL
Fringilla sp.	FRI
Fringillidae	FEE
Hirundinidae	HEE
Loxia sp.	LOX
Motacilla sp.	MOT
Parus sp.	PAR
Passer sp.	PAS
Passeriformes	PSS
Streptopelia sp.	STR
Turdus iliacus/philomelos	TUR

6.2 Sex/Age Determination and Coding

In modern ringing, sexing and ageing of ringed birds is a rule. However, contrary to the species determination, it is sometimes impossible to determine, or the criteria do not allow full separation. Despite this weakness, sex and age determination should be performed in as many cases as possible. Where possible, species-specific characters should be used according to appropriate manuals. As in the case of species determination, key or guide procedures could be used. In the cases of sex and age determination, where different characters are more diversified in terms of their utility and simplicity to apply, the key system has some additional advantages. It is a psychological rule, that beginners use their own intuition, which could be far from optimal, in sex and age identification, especially when confronted with a few unprecedented characters. Since not everyone knows these solutions, the results of identifications made by different ringers may have different degrees of validity. They depend on the leading character accepted by the ringer and need not be fully compatible. On the other hand, a hierarchy of characters instituted by an experienced specialist and presented as a key hierarchy has a chance of being optimal.

Since most age characters are connected with the bird plumage, the age coding is derived from the names of subsequent plumages, which are identifiable (Figure 6.1):

J – juvenile plumage: the bird in its first full plumage; feathers grown in the nest or first feathers following down-plumage

I – mixed immature plumage: plumage containing some juvenile feathers and some renewed feathers of the next set

A – definite (adult) nuptial plumage, **D** – adult postnuptial plumage

N – not defined plumage, but not juvenile plumage

L or "**-**" – full-grown, not checked for plumage type. For some species with more complicated pattern of plumage development, a few other, detailed codes could be used:

S – first full immature plumage: second full plumage

T – second full immature plumage: third full plumage

M – immature plumage, precise type (I, S, T) is unknown

O – not adult plumage (either juvenile or any immature one)

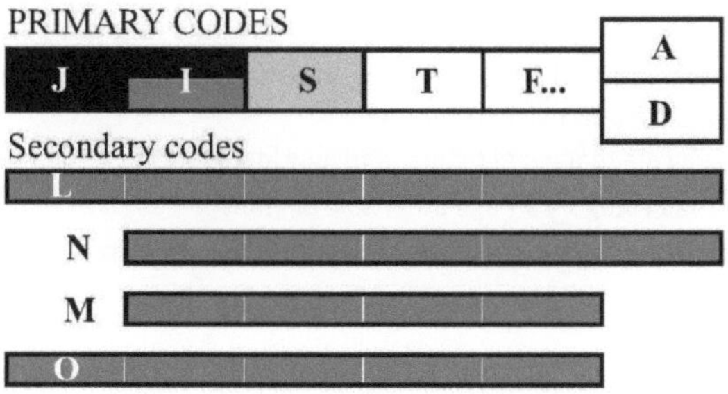

Figure 6.1: Plumage coding system (explanations in the text). Secondary codes are used when the exact plumage is not known. They could contain different combination of plumage classes as shown at the bottom of the Figure.

Apart of the age coding system based on the plumage (codes given above), a system called "calendar years" is also used by ringers. In this system, a young bird (from this year's brood) is called "1st year bird", while its parents are in "after 1st year". It seems simple, but the first year lasts only till December 31st, so has a length of few months (in Europe), while it can last a few days only in the tropics and on the southern hemisphere, where breeding time is around December/January. From the 1st of January the same bird called earlier as "1st year" must be described as being in "2nd year" and its parents as "after 2nd year". The next change of the age name is when birds leave nests, our "2nd year" bird is named as "after 1st year", while its parents become apparently younger, written as "after 1st year" instead of "after 2nd year". The next complication due to the differential coding of "ages" in different countries: as number codes (EURING, Great Britain) or after local languages (e.g. "1", "2", "po 1", "po 2" in Poland or 2, 4, 5 and 6 in Italy). The codes used in official EURING system are as shown in the Table 6.3.

Table 6.3: EURING age codes.

EURING code	
0	Age unknown, i.e. not recorded.
1	Pullus: nestling or chick, unable to fly freely, still able to be caught by hand.
2	Full-grown: able to fly freely but age otherwise unknown.
3	First-year: full-grown bird hatched in the breeding season of this calendar year.
4	After first-year: full-grown bird hatched before current calendar year; year of birth otherwise unknown.
5	2nd year: a bird hatched previous calendar year and now in its second calendar year.
6	After 2nd year: full-grown bird hatched before previous calendar year; year of birth otherwise unknown.
7	3rd year: a bird hatched two calendar years before, and now in its third calendar year.
8	After 3rd year: a full-grown bird hatched more than three calendar years ago (including present year as one); year of birth otherwise unknown.
9	4th year: a bird hatched three calendar years before, and now in its fourth calendar year.
A	After 4th year: a bird older than category 9 - age otherwise unknown.
B	5th year: one year older than category 9 - age known exactly to the year. C, D, E, F, G, H, J, K, L, etc. onwards *et seq.*

In the sex coding only letters M (male) and F (female) or scientific signs (♂ and ♀) should be used – any number coding (0 and 1, 1 and 2, 2 and 1) easily leads to errors!

6.3 Standard Set of Measurements

First, one general comment: It must be emphasized that a single measurement of a ringed individual is of very low value to the study of population differentiation among migrants, their breeding origin, or even of sex and age dimorphism. Despite common beliefs, a measurement of e.g. only wing-length certainly does not present adequate information on "the size of the bird". Different measurable bird size parameters including wing-length, feather-length, tail-length and body mass are sometimes not positively correlated when birds of different origin are compared. These parameters often change independently over the breeding range, so we may be confronted with e.g. long-winged and short-tailed birds at one station, and short-winged and long-tailed at another. At the same time, body mass depends very much on fat reserves of an individual, and heavier, but fatty, birds may still have lower lean body mass than other individuals, which were weighed as lean birds. Arrangement of measurements into carefully selected standard sets allows us to conduct much more detailed biometric studies. These sets may be different in e.g. passerines and waders.

A recommended standard set of measurements for passerines contains wing- and tail-length, wing formula, fat score and body mass. For standards applicable to waders see, *Wader Station Laboratory Methods* (p. 148).

6.4 Standard Descriptions of Measurements

Note that standard bird handling (described earlier) is assumed for all of the procedures recommended here. The ruler of 30 cm long and cut off at the zero-line is the standard tool for wing, wing-formula and tail measurements.

Within descriptions of alternative methods (see *Alternative Methods of Holding and Measuring Birds*, p. 184) the manner of handling could be different, and other types of rulers could be used.

Wing-length (maximum chord measurement)

Both the ruler with and without a stop can be used, as they give exactly the same results.

Technique

The folded wing, parallel to the body axis, rests on a ruler (Figure 6.2-6.3). The carpal joint of the wing is placed at the butt of the ruler; if the ruler has no stop, cut off at the zero-point, using the bulb of the second finger of the right hand. With the thumb of the same hand, the wing is firmly but carefully pressed against the ruler; at the same time, the thumb of the left hand straightens the primaries to their maximum length by smoothing the lateral curvature and applying slight lateral pressure towards the bird's body at the level of the primary coverts. The third and fourth fingers of the left hand control folding and straightening of the wing.

Precision of measurement: 1 mm

The most common mistakes in measurement:

Wing-length underestimated:

- The wing not fully pressed to the ruler, the primaries not fully straightened.

Wing-length overestimated:

- The carpal joint not taken fully to the butt of the ruler (this is much more probable when a ruler with a stop is used).
- The carpal joint is not bent enough, and bones of forearm are included in measurement.

Other techniques – see *Alternative Methods of Holding and Measuring Birds*, p. 184.

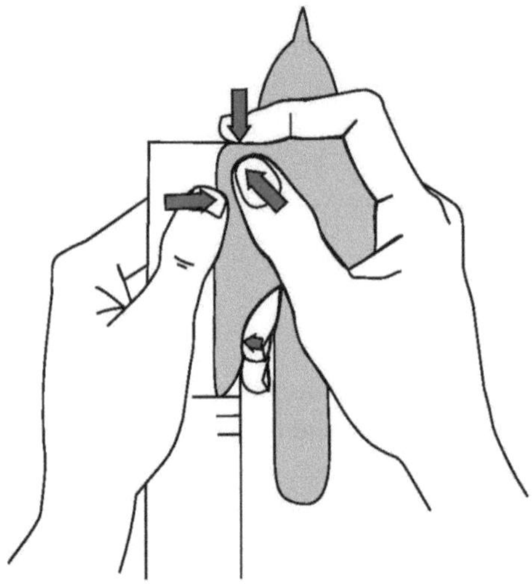

Figure 6.2: Standard measurement of the wing-length (critical elements of the procedure pointed by arrows).

Wing-formula measurement (feather tips distances method)

The wing-formula includes measurements of distances from the wing point to the tips of the shorter primaries. The primaries from the second to the eighth (in ascending order) are taken into consideration (Figure 6.3); for simplicity, the first functional primary is always numbered as second, irrespective of its "evolutionary" number (even in families e.g. *Motacillidae, Fringillidae,* etc., which have lost their first short primary).

Use of a ruler with a zero-stop is not convenient, though possible.

An example wing (Figure 6.4): the tip of the wing is formed by primaries 4th and 5th; 3rd primary tip is by 1 mm shorter than the wing-tip, 6th – 2 mm, 2nd is equal to the 7th and they are shorter by 6 mm, 8th is 9 mm shorter.

For recording purposes, this formula would be spoken as: "four to fifth, zero-one, two, six-six, nine". The record in subsequent boxes of the form:

| 45 | 01 | 2 | 6 | 6 | 9 | |

"Zero" (in box 2) is written as an indicator for special processing.

Explanations of spoken recording:

1. In the first box – the numbers of the longest primaries are called out,

2. In the next boxes – distances (in full mm) between the tips of primaries and the tip of the wing.

(1) "fourth to fifth" indicates that primaries 4 and 5 form the tip of the wing (4=5). Other possibilities in this box: (A.) only one number (e.g. "third" – means that the tip of the wing is formed by one primary only (the third); (B.) two not consecutive numbers (e.g. "fourth to sixth" means 4=5=6, "third to sixth" - 3=4=5=6).

Figure 6.3-1: Measurement of the wing-length taken using a ruler **without** stop. Photo unknown.

Figure 6.3-2: Measurement of the wing-length taken using a ruler **with** stop. Photo unknown.

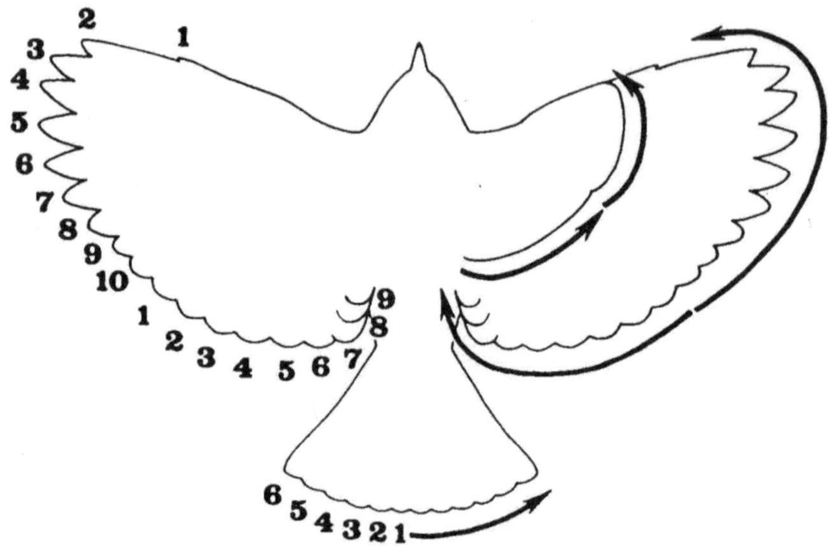

Figure 6.4: Ascendant numbering of primaries and rectrices. Typical moult directions are shown at the right side of the drawing.

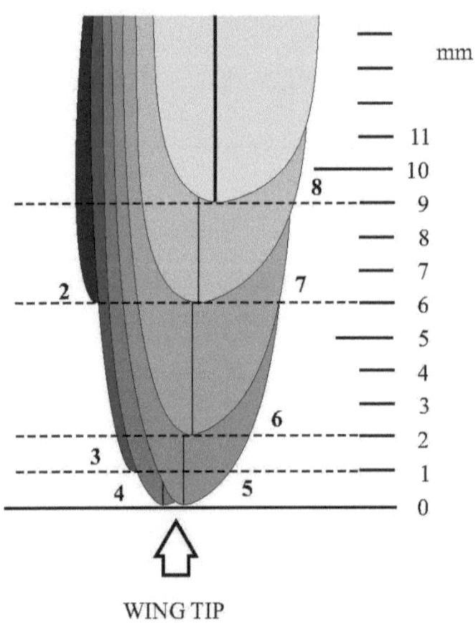

Figure 6.5: The wing-formula measurement (example used in text). Note that this is NOT an illustration of the technique of the measurement (see Figure 6.7).

(2) "zero-one" – the word "zero" indicates that the measurement given is of the distal primary (i.e. placed distally in relation to the longest ones; in this formula the "distals" are the second and third primaries and the "proximals" – 6th, 7th and 8th; when the distal primary is equal with the proximal one, the word "zero" is omitted (e.g. "six-six" – later in this formula).
(3) "two" and "nine" – the measurements of the proximal primaries.

Note: if someone would like to measure all primaries – till the 10th – two more boxes should be added into the layout shown above.

Technique (Figure 6.6)

1. Count the number(s) of the longest primary(ies) when the wing is closed before measuring spread and extending the left wing to check the state of the feathers for cleanliness, moult, loss or damage (and to check that they follow in correct sequence).

2. Fix the closed wing in its natural position (as natural as possible), holding it almost parallel to the body axis (looking from the back side) by holding with the first and second fingers of the right hand near the carpal joint so that the primaries cannot change their position during the measurement procedure.

This is the most difficult and critical part of the technique.

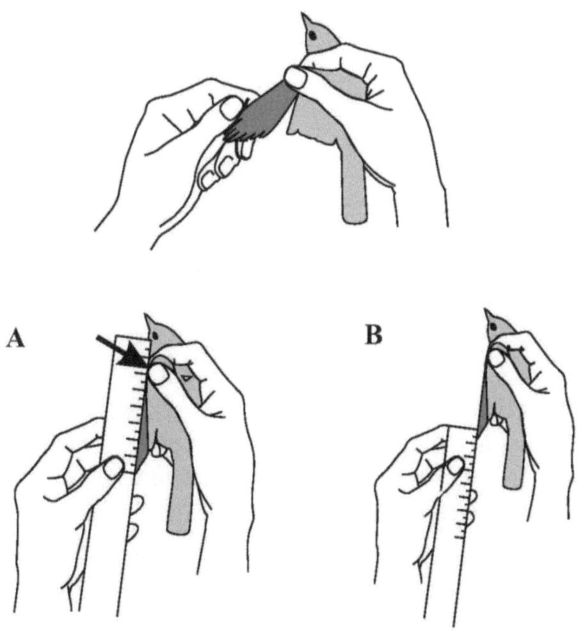

Figure 6.6: Technique of the wing-formula measurement. Clearing the feathers sequence (above) and two variants (A and B – see text) are shown (below).

3. Measure the wing-formula with the methods A or B (see below). To measure the distal primaries move the hand with the bird in relation to the fixed ruler position.

A: the edge (the zero-end) of the ruler is placed at the tip of each primary sequentially from the wing tip. After each value has been recorded, the ruler's end is moved to the next primary tip.

B: the tip of the wing is put at any centimetre-line of the ruler (convenient for the size of the bird) and the values are taken in the opposite direction from that normally used.

These two methods are compatible, but Method A is slower and more vulnerable to mis-readings. Thus, Method B is generally recommended. If the same person is measuring wing-formula and recoding it into a notebook, it is highly advisable to measure and memorize the whole formula at once and then write it at once, too.

Precision of measurements: 1 mm

The most common mistakes in measuring:

Mistakes mainly result from inaccurate handling of the bird:

- The head of the bird is pulled back between fingers, and
- The wing is extended too much and not firmly fixed between the first and second fingers.

When method B is used, the position of the wing tip on the ruler may change, causing inaccurate measurements if the ruler is not fixed (in relation to the hand holding the bird).

Comments to the other method – see *Alternative Methods of Holding and Measuring Birds*, p. 184

Tail-length measurement ("to the back" method after Busse, 1983; 1990)

Measurement of the tail with the pygostyl: a simple, very quick and safe method for the bird. The ruler with the stop cannot be used.

Technique

The body of the bird is held **vertically** with the tail directed at right angle (90°) to the back (Figure 6.7). The ruler lies at the tail with the butt pressed firmly to the back (controlled with the fourth and the fifth finger under the belly). **The restrices should touch the ruler along their whole length.** The longest tail feather measurement is read.

Generally, tail length measurements are more difficult to be taken correctly, so special attention should be put on bolded remarks in the description.

Precision of measurement: 1 mm

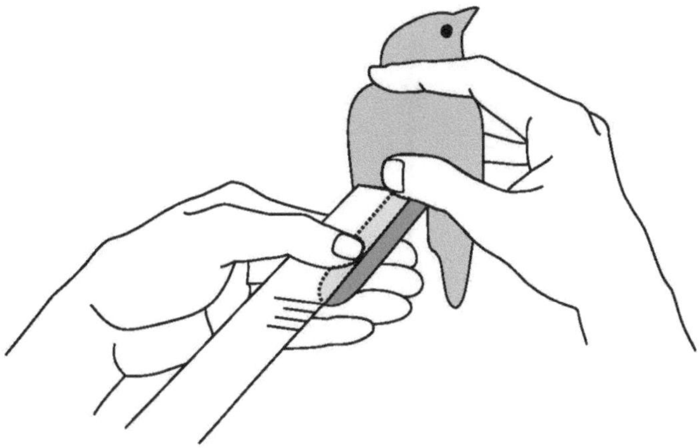

Figure 6.7: Technique of the tail-length measurement (note perpendicular position of the birds back in relation to the ruler).

The most common mistakes in measurement:

Tail-length underestimated:

- An acute angle made between the tail and the back; the bird body too close to the ruler).
- The ruler end not firmly located (the fourth and the fifth fingers do not press the belly).
- Restrices do not touch the ruler along their whole length (not pressed to the ruler by the fingers of the left hand).
- False reading of some mm will result if the butt of the ruler is at the tip of the pygostyl instead of at the bird's back.

Tail-length overestimated:

- An obtuse angle made between the tail and the back; the bird body too far from the ruler.
- The butt of the ruler touches the bird's back well above the pygostyl instead of laying on it.

Other techniques – see *Alternative Methods of Holding and Measuring Birds*, p. 184.

Fat determination (after Busse, 1983 and Kaiser, 1993, combined)

Determination of fat goes through three levels (Figure 6.8):

Level I – belly
Level II – furculum
Level III – pectoral muscles

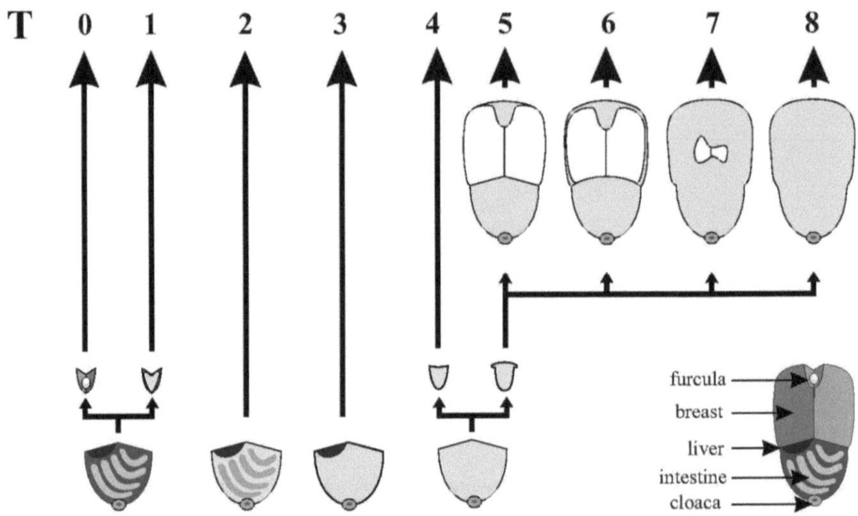

Figure 6.8: Fat scores (description in the fat-scoring key on p. 95).

Key to fat determination:

I. 1. Belly is without visible fat or with reddish traces only – **II A**
2. Belly with infused bands of fat (intestinum is visible) **T2**
3. Belly has a fused cover of fat; intestinum is not but the liver is visible **T3**
4. Belly is completely covered with fat, a very narrow band of the liver may be visible but, if this is so, the roll of fat is just above it – **II B**

II A. 1. Air -sack is visible within furculum (some fat may occur) **T0**
2. All the interior of furculum is covered with fat **T1**

II B. 1. Fat in furculum flat or concave **T4**
2. Fat in furculum forms a convex cushion – **III**

III. 1. Sides of pectoral muscles without strips of fat **T5**
2. Sides of pectoral muscles with strips of fat **T6**
3. Pectoral muscles partly covered with fat **T7**
4. Pectoral muscles completely covered with fat **T8**

Note: In some species, loss of fat does not follow exactly the same sequence in which it was attained, this results in problem with fat determination in some specimens. Anyhow, always follow exactly the key, as specific differences are covered by species-specific validation of the scale.

Technique:

1. Lay the bird on its back on the palm (Figure 6.8); the neck should be between the second and third fingers of the hand; the second and the third fingers of the second hand should gently part the bird's legs; the proper position of bird is very important.
2. Blow the belly (Figure 6.8A) with a continuous stream of air and choose one of four possibilities under section I of the key; if the second or third subsection is chosen you have determined the fatness as T2 or T3 respectively.
3. If II A or II B are chosen, you must direct your blowing to the furculum (Figure 6.9B) and choose one of the two subsections under II A (fatness T0 or T1) or II B (fatness T4 or higher – III).
4. If your choice is III, look at pectoral muscles and choose fatness T5 - T8.

The most common mistakes:

Mistakes are usually made when someone has a tendency to take "liberal" interpretation of rules, e.g. when the bird has a thick cover of yellow fat on the belly but part of the intestinum visible; this should be T2 but is classified as T3 because it "looked like a fatty bird". Some mistakes are possible when the bird is not properly handled when the furculum contents are evaluated.

Note that the fatness of an individual bird properly determined twice at the same time may be different. This is because in border cases, different tension of the bird's belly muscles at the moment of blowing may expose (or not expose) the intestinum or the liver from under the fat layer. Difference in determination cannot, however, exceed one degree of fatness.

A B

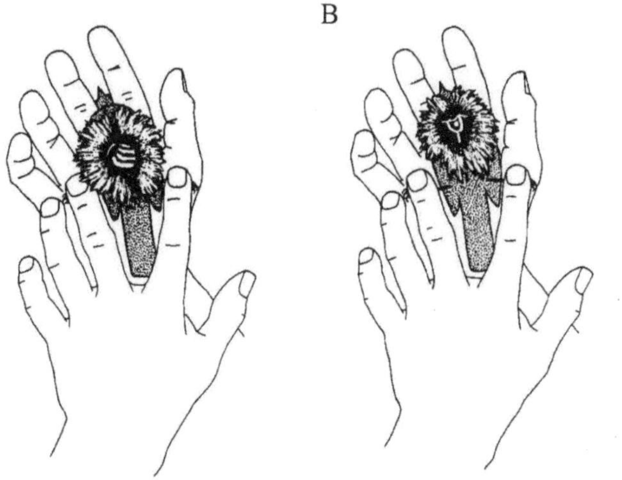

Figure 6.9: Technique of the fat scoring. A – blowing to the belly, B – blowing to the furcular depression.

6.5 Additional Measurements and Scores

Additional measurements and scores can be used optionally.

Feather-length of the third outermost primary

Description from *Manual of Field Methods* (Bairlein, 1995): "Measuring feather-length takes little time and can be accurately done when the following instructions are observed (Figure 6.10):

Use a ruler with vertical pin of exactly 1.4 mm diameter (Figure 3.7 (3)).

The ruler has to be fixed onto a block of wood or onto the table and the bird has to be held with both hands. Do not hold the ruler free-hand. By using this method the inter-observer variance of the measurement is significantly reduced.

Hold the wing at the carpal joint between your thumb and index finger. Take the second outermost primary (F9) with the other hand and open the wing slightly and place the pin between 2nd and 3rd outermost primaries until it firmly touches the skin. This point is easily found and well defined.

The primary now has to be completely straightened by first bending it outward a little (to get maximum length). The length is read to the nearest 0.5 mm.

Make sure not to interfere with primary coverts, i.e. the primary covert should be on the same side of the pin as the corresponding primary.

Do not use excessive force, and be as cautious as possible to avoid any injuries."

Comments: The method presented, although it seems very exact, has a number of disadvantages. First of all, the first measurement cannot be repeated on the same individual as the second, and subsequent measurements are regularly 0.5 to 1 mm longer. These are corrected for during calibration courses in which trained ringers participate; this means that there is no way to control for whether the newly trained person measures

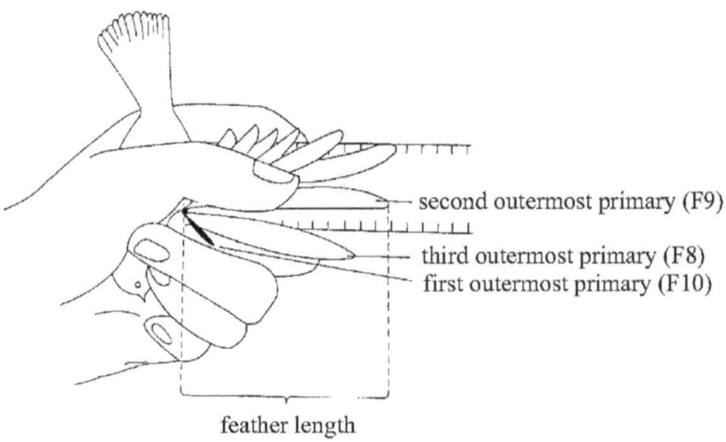

Figure 6.10: Technique of the feather-length (after Vogelwarte Radolfzell, from Bairlein, 1995).

the bird correctly or not. Secondly, the diameter of the pin (1.4 mm) is much too thick to measure the feather-length in small passerines (a distance between primaries at the level where they go out from the skin is around 0.2 mm). Thus, there remains the possibility to use the excessive power to press the pin until it firmly touches the skin, and when a 1.4 mm thick pin is pressed into a fissure that is several times narrower, the skin may be broken or slipped along the feather shaft. Subsequent measurements show that this is not only a theoretical speculation. Discussion of the method by Gosler *et al.*, (1995) shows that it should not replace wing-length as a standard measurement.

Wing-shape measurement

The wing-shape measurement as described in *Manual of Field Methods* (Bairlein 1995): "To measure wing-shape, the length of each individual primary (except the outermost F10) and the first secondary is measured using the feather-length ruler and the method described above (feather-length measurement). With the exception of primary 9 (F9, the second outermost), which has to be measured with the pin inserted between F9 and F8, the pin has to be inserted on the "outer" side (distally) of each primary/secondary to be measured (Figure 6.11). For wing-shape, read feather length to the nearest 0.5 mm.

It does not matter how the ringer holds the bird and which wing is measured."

For more detailed description of the method see feather-length measurement description.

Comments: Firstly, as stressed previously in the description above, this is not the wing-shape measurement but rather a set of ten independent measurements of ten bird feathers. This is easily done and the result, obviously, may be elaborated on according to specific needs. However, the lengths of subsequent primaries do not describe the wing-shape, as they are located at different angles and in different places along the carpal part of the wing, so the real wing-shape is derived from both, (1)

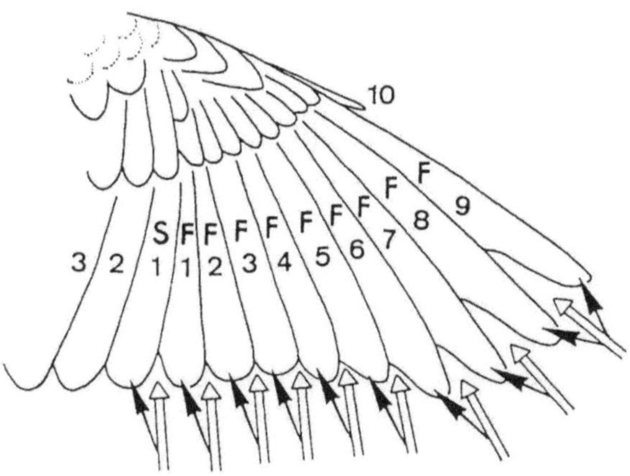

Figure 6.11: Idea of the wing-shape measuring (after Jenni and Winkler 1989, from Bairlein, 1995).

lengths of subsequent feathers and (2) peculiar features in their distribution along the wing. Secondly, the measurement of the outer primary is taken from the other side of the feather, which means that it is not comparable with the other measurements. The primaries extend in a step-like fashion along the wing so, the measurement "from below" is not equal to the measurement "from above" of the feather (i.e. the edge of the feather that points rostral vs. caudal). Apart from writing comments to the feather-length measurement (which should also be done here), this method is extremely time consuming, so in practice, it is not useful when ringing a lot of birds.

Bill-length measurement

The usefulness of bill-length measurement differs very much in various groups of birds. This is a standard (and very useful) measurement in waders, but it is of a very limited value in passerines. It can be done using callipers or dividers as shown at Figure 6.12.

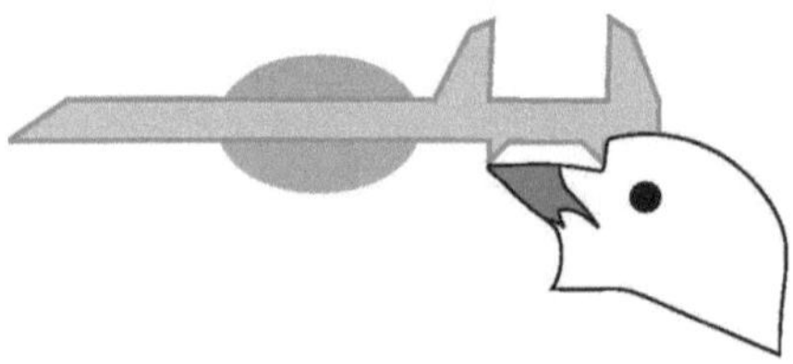

Figure 6.12: Measurement of the bill-length to the skull in passerines.

Tarsus-length measurement

The usefulness of tarsus-length in biometrical studies also differs greatly, and although it may be useful in wader studies, nobody has shown the same for migrating passerines. However, there are papers showing the tarsus-length to be useful as an additional measurement for certain species. For example, it is used as an index of a body size in special studies on breeding populations.

In passerines, two methods are used.

Measuring with dividers, as given in *The Ringer's Manual* (Spencer, 1972): "The measure is taken from the depression in the angle of the inter-tarsal joint (the "knee") to the base of the last complete scale before the toes diverge (Figure 6.13A). It is the length of the tarso-metatarsal bone that is measured. It is recommended that the tarsus should normally be measured to the nearest 0.5 mm, but to the nearest 1 mm in species with tarsi measuring 60 mm or longer."

Measuring using callipers is presented in *Manual of Field Methods* (Figure 6.13B): "The following instructions for measuring tarsus are for a right handed person. For a left handed person: reverse left and right hands. The position of the right leg of the bird will be somewhat different. Use easily running slide callipers, and be careful not to bend the tibiotarsus.

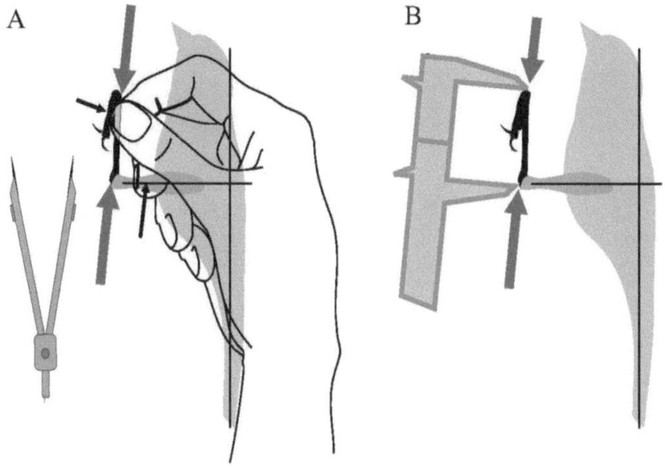

Figure 6.13: Two methods of the tarsus-length measurement. A – using dividers, B – using callipers.

1. Take the bird, lying on its back, in your left hand with the bird's head between your index finger and your middle finger.

2. Hold the right (meta)tarsus between thumb and index-finger, fold the toes backwards and also hold them between thumb and index-finger. For birds with a very short tarsus one should use the extreme tips of the fingers.

3. Position the tip of the middle finger behind the tibiotarsus, such that the tibiotarsus makes a right angle to the body and the metatarsus makes a right angle to the upper part of the leg. This positioning greatly improves the within and between observer repeatability of the measurement.

4. Make the measurement from the notch on the metatarsus to the top of the bone above the folded toes (Figure 6.13B), and read the callipers to 0.1 mm".

Muscle-score

Description after *Manual of Field Methods* (Barlein, 1995) (Figure 6.14): "Beside fat, which is the primary energy fuel for migrating birds, migrants also use muscle proteins in flight. The size of the breast muscle is a further valuable indicator to body condition of migrants. In birds whose flight muscles are not covered by fat, the shape of the breast muscles can be easily recorded and scored. Muscle score is assessed visually and by sweeping the thumb over the sternum".

Comments: It seems that this scoring is useful for the birds of low (or very low) fat reserves. In assessing the muscle-score, one must remember that there are distinct and specific differences pertaining to breast muscle appearance: some species have nearly always a "good look" while others always have a "slim" appearance.

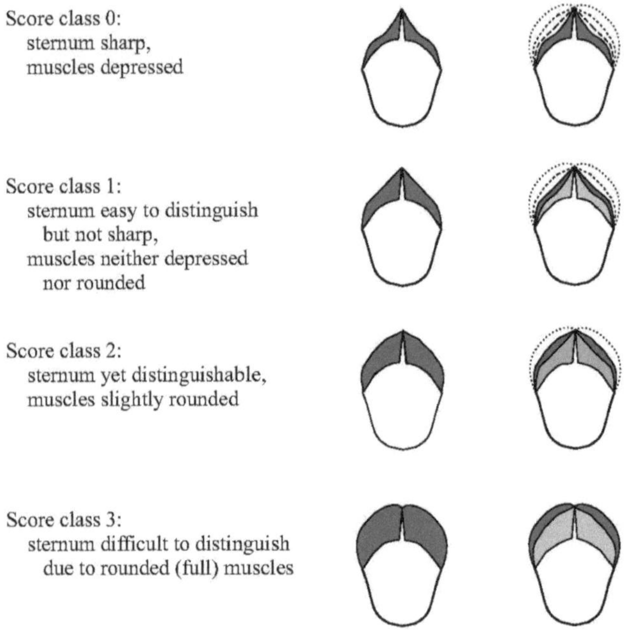

Figure 6.14: Muscle scoring (after G. Wallinger from Bairlein, 1995 modified).

6.6 Special Studies

6.6.1 Directional Preferences of Nocturnal Migrants

The described method of studying directional preferences in nocturnal migrants includes a new field technique (Busse's flat cage – Busse, 1995) and pays special attention to the inconsistency of directional behaviour pattern in individual birds. It may be used under real field conditions, by professionals as well as amateurs. The equipment is simple and cheap, and the technique easy to learn in a standardized format. Additionally, the test routine allows for the collection of large amounts of data (since tests may be performed both at night and at day). Diurnal tests under an overcast sky have the same value as tests done with good sky visibility, but this is not the case in nocturnal tests. Analysis of local vectors of directional behaviour patterns seems useful in studies on local migratory directions and on the overall population composition of migrants.

General idea of the flat cage. A bird tested in the flat cage inside of the screen is cut off any visual cues but the sky only. The protection screen (see Figure 6.15 for dimensions) idea was shown in Figure 3.9 (screen partly removed on the picture to show the cage inside).

The testing stand. The place of tests should be a flat area (top of a hill, etc.) without trees, wires, or poles that may be seen by a bird above the protecting screen (Figure 3.10.)

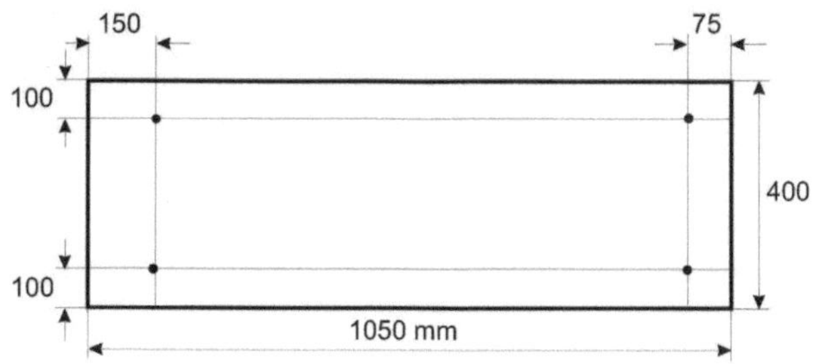

Figure 6.15: Dimensions of the one of four segments building the protective wall at the orientation tests stand.

The testing routine. Tests can be done at any time, both night and day. There are meteorological limitations, so tests should not be done with rainfall or snow (nor with wet fog) causing condensation on the foil of the test cage. Tests are not recommended with wind force exceeding 5° Beaufort.

Caught birds can be tested immediately after catching and ringing, or they can be kept in opaque bags or cages for at most two hours.

1. Preparation of the cage for the test includes covering its vertical sidewall with a strip of foil from a roll of width adjusted to the height of the cage (with an extra 2 cm for folding – Figure 6.16). Details of the procedure are given in Figures 6.17-6.18. Fix the beginning of the strip to one of vertical wires of the cage with transparent sticky tape, and then cover the side of the cage with straightened foil, fixing its upper end. Finally, cut the strip off of the roll after connecting the ends. The foil should be carefully handled to avoid making scratches, holes etc.

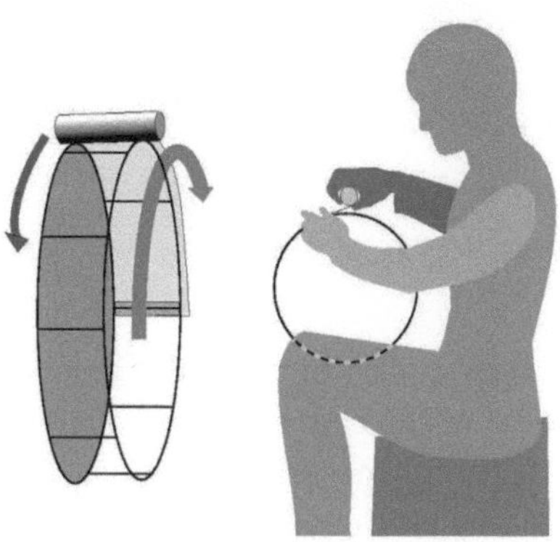

Figure 6.16: Putting-on foil on orientation tests cage – this is the only effective position for a person performing this task.

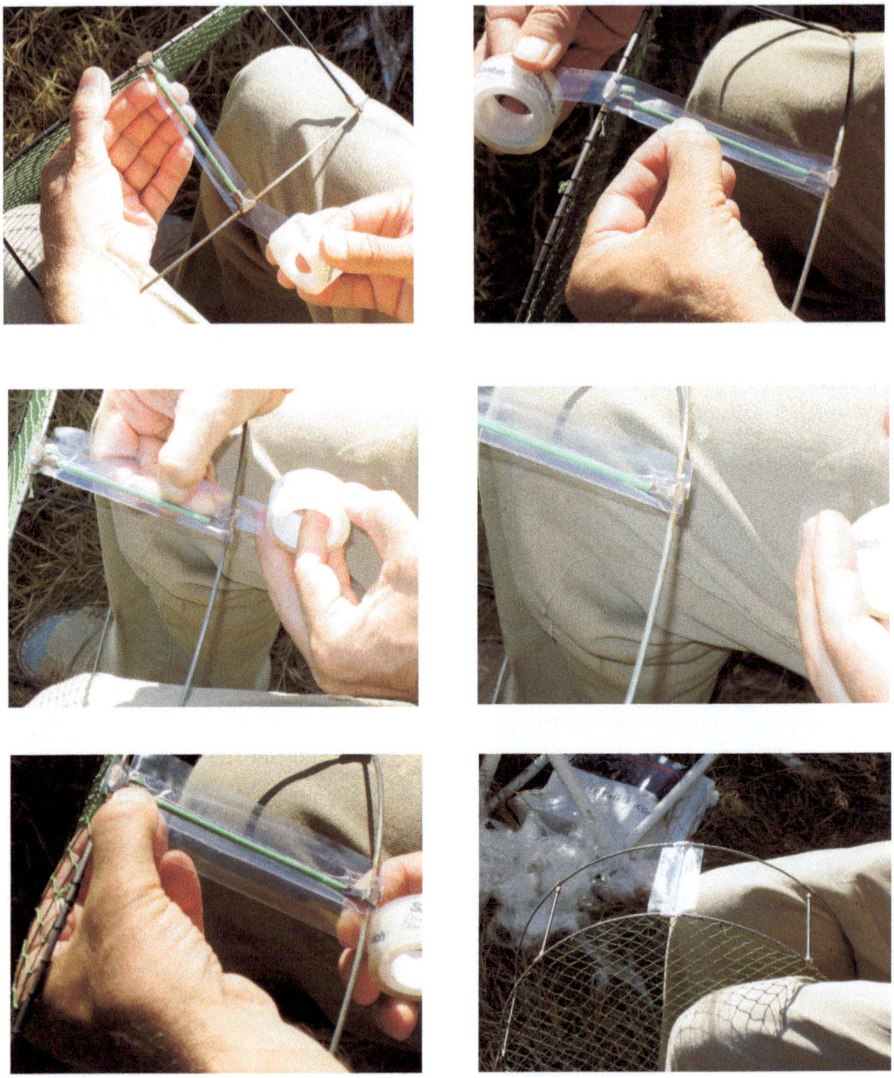

Figure 6.17: Preparing a cage for the test – first step (creating the starting stick; description in the text).

(which could subsequently be taken for the signs of bird activity). The cage may be prepared in advance, but extended storage of the cages in moist air is not recommended since the sticky tape used to fix the foil may come loose.

2. Transport the bird to the test stand in an opaque bag or cage, remove it below a top of the protecting screen and put it into the test cage inside the screen (Figure 6.19-1). The direction the bird enters the cage does not seem to influence the results, but the custom of putting it from one side (e.g. always from the south) could be a rule.

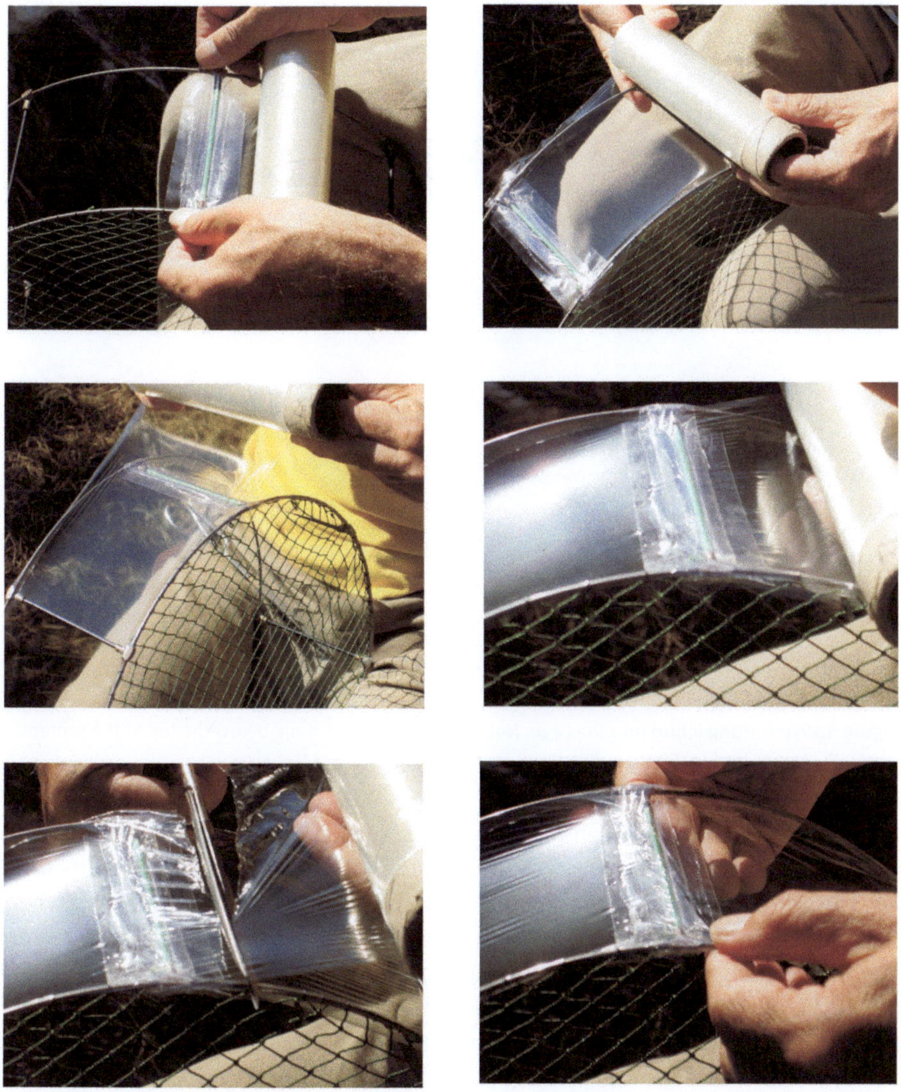

Figure 6.18: Preparing a cage for the test – second step (covering the side wall with foil; exact explanation in the text).

After inserting the bird, the observer should leave the place quickly and note the time (precision 1 minute), and after the agreed test time (10 minutes proposed as standard), quickly return and remove the bird from the cage. If the bird is earmarked for other tests, it must be caught by hand (which is not too easy and many escape). During the test time, the bird should not be disturbed by sudden noises or things coming into visibility. When larger birds as thrushes are tested, the cage should be fixed to the ground so fluttering does not displace the cage.

Figure 6.19-1: Putting a bird into the cage. Note that operation is done below the top of the protecting wall. Azraq, Jordan. Photo P. Busse.

3. The test cage is at the centre of the protecting screen (Figure 6.19-2) with one of the wires directed to the North (indicated by a previously fixed pole outside of the screen, not visible to the bird). It is always handy to direct the wire where the foil strip is fixed (to the North); this prevents misidentification of sectors when noting the results. Putting the cage on a flat, symmetrical piece of carton without a slimy surface is highly recommended.
4. After the test, the results of the test should be noted. Count signs of the bird activity sector by sector. Starting always from NNE direction is convenient when you handle the cage (with its bottom side to your belly, Figure 6.20-1). Count the signs of activity: holes and dots made by bill as well as holes and scratches made by claws of bird when it hopped against the foil. Sometimes, signs of different origins are not easy to separate, so counting them altogether is the best solution. The behaviour of the bird in a cage is, to some extent, species-specific, and in one species, bill signs are more common, while claw marks may be the majority in another. Some practice is needed, but individual differences between observers, if they do exist, concern the number of counted signs and not their distribution. Every counted sign must be instantly marked with colour marker to avoid double counts. It is a good custom to write numbers on the foil first and then rewrite them into the form.

Figure 6.19-2: A bird in the cage. Note bottom plate and details of the wall construction. Burullus, Egypt. Photo P. Busse.

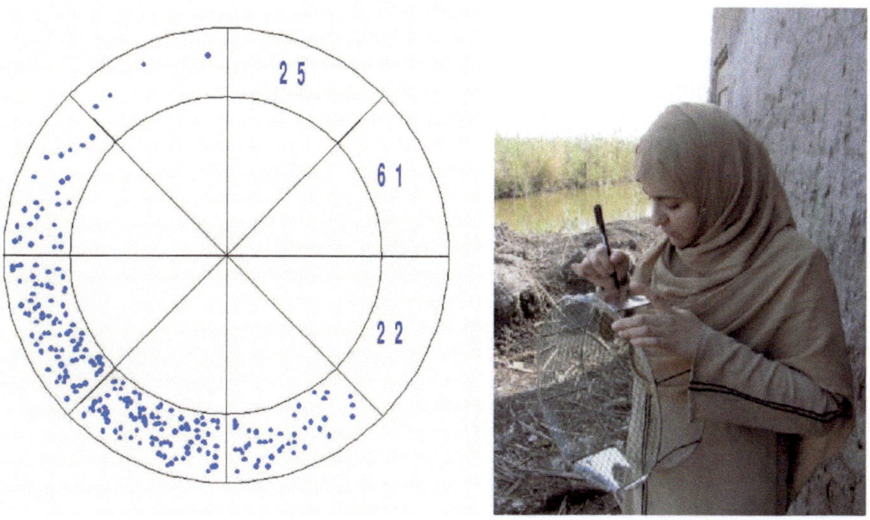

Figure 6.20-1: Counting the scratches on the foil. However, an advice is to do this sitting. Burullus, Egypt. Photo P. Busse.

Figure 6.20-2: A special note-book with results of eight tests.

Note that, if you handle the cage as recommended above, the correct direction of writing into the circular data form is opposite (Figure 6.21).

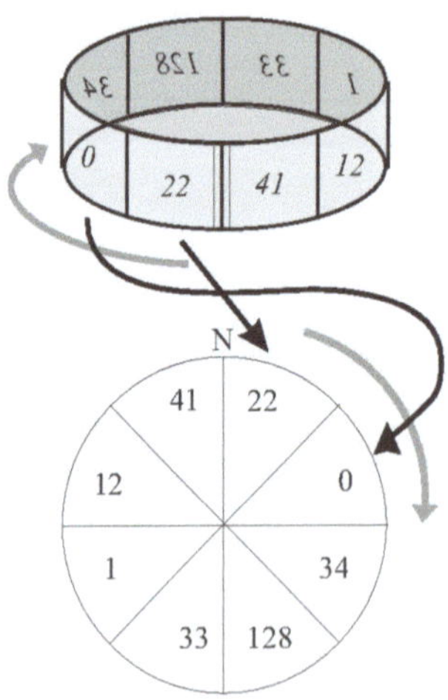

Figure 6.21: Noting the results of the orientation test.

Longer storage of cages before counting is not recommended because one may risk accidental damage of the foil. However, as it is easier and quicker to count the signs made by the bird in good light conditions, cages from the night tests may at least be stored till next morning, if there are enough cages for all planned tests. Used foils cannot be handled or stored after removing them from the cage.

5. Filling out the test form includes a couple of boxes with information complementary to the main data (Figure 6.22):

   Species, Ring number, Status (A - freshly ringed, first test, B - next test; R - retrap), Sex/age, Fatness (the fat-scale used is specified separately), Date - hour of catching, test time (from - to, given as hour and minutes), Day/night (D, N), Sky visibility (0 - none, 1 - small: cloudiness 7 to 9, 2 - medium: 4 to 6, 3 - good: 0 to 3), Sun/Moon (S - the Sun, M - the Moon visible, „-" none of them), Wind direction (accuracy to 1/8 of the wind-star; 0 - no wind), Wind force (0 - no wind, 1 – 1° to 2° Beaufort, 2 – 3° to 4° B, 3 - over 4° B).

Species		
Ring		
Status		
Sex/Age		
Fat-Weight		
Date-hour		
Experiment	from	to
Day/Night		
Sky		
Sun/Moon		
Wind dir.		
Wind force		

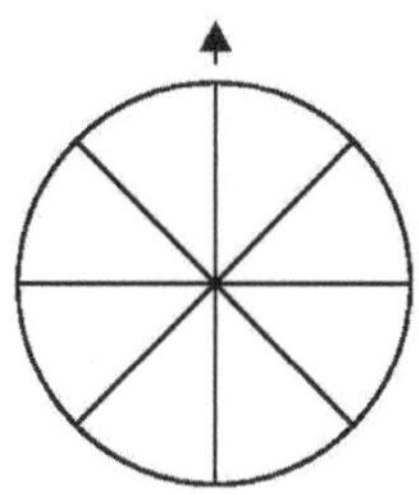

Figure 6.22: Orientation tests – the data form.

After filling up the test form, the foil is removed from the cage, and the cage may be prepared for the next test. One single person working at one test stand may, without problems, handle six birds per hour, including the need to count results and prepare the next cages, as long as the test stand is not too far from the ringing site. Working two stands is possible for an experienced person, otherwise, it requires some help from a second person.

6.6.2 Additional tips

1. *Building up the screen*

A. The screen is built from four pieces of solid elastic plates. Every part of the screen is identical, with four holes for screwing them together (Figure 6.15).

Note, however, that at one end, holes are farther (end A) from the shorter side of the plate than at the second end (end B).

B. For preparing parts for connecting - lay them in one row with ends A laying ON ends B. The upper part of this strip will be inside of the screen. Connect parts, using screws, with their heads inside of the screen. We suggest using thumb-nuts. This makes a long strip of four parts together.

C. Connect the beginning with the end of the stripe using screws in the same manner: end A of the last part must be INSIDE the round screen.

2. *Preparing the cage*

Put rods into holes in a circular cover of a cage (this with a net) and fix them by pressing or beating using a piece of wood. Put circular base on rods and fix them using the piece of wood. For longer transport, you can remove rods.

3. *Covering the cage with a foil*

A. Note that one of the rods is a different colour from than the rest. Use it as the beginning and the end of the foil, for fixing see point 3B, it will be located North!

B. Use transparent sticky tape about 20 mm wide:

1. Start fixing this tape from top end of the rod *inside* of the cage but with sticky side *outwards* (the rod must be placed symmetrically); the tape should go around base circle wire and return to the top of the rod: both inner and outer parts are fixed together making a base for fixing "wings".
2. Make "wings" by sticking the tape from *inside* on both sides of the base: you have two sticky wings along the rod.
3. Fix the cage between laps as on Figure 5.4, fix the beginning of a foil stripe to one of wings, cover all sidewalls with strengthened foil and fix it to the second wing. Cut off the foil, but with some margin to cover second wing too.

6.6.3 The Study of Moult

Moulting strategies and the timing of moult are highly varied, depending on the population. Since various populations migrate all over Europe and the Middle East, records of moult in migrants offer many possibilities for interesting moult studies. One course of action is to collect moult data using "moult cards". The moult card design for passerines and its filling-up rules are presented below (Figure 6.23) after instructions of the Swiss Ornithological Institute:

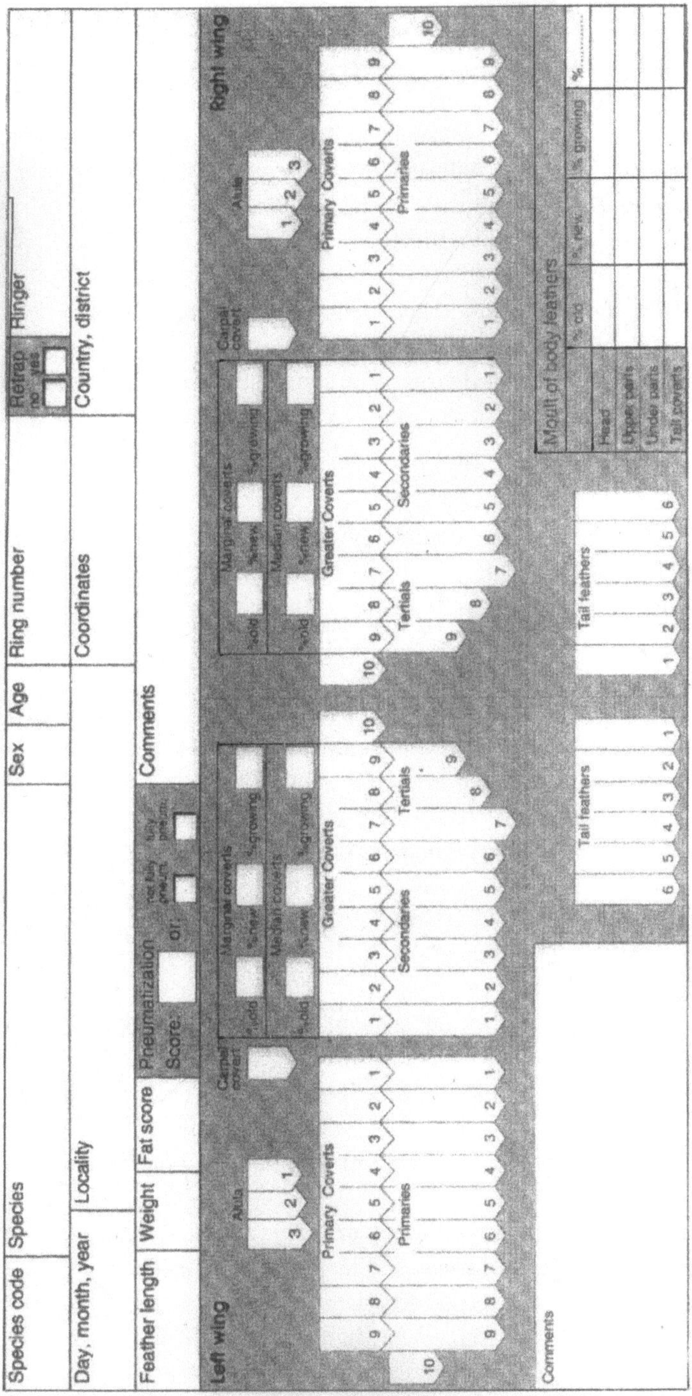

Figure 6.23: Moult card of the Swiss Ornithological Institute.

"This moult card shows both wings. For general use, fill in moult cards for every bird belonging to one of following categories:

1. In summer/autumn: All adults in moult and all adults having renewed part of the plumage; all first-year birds with growing or renewed secondaries or primaries.

2. In winter/spring: All birds in active moult of primaries, secondaries, tertiaries, rectrices or greater coverts; all birds after moult with renewed secondaries or primaries.

If possible, record all feathers, but data for restricted tracts only (e.g. primaries and secondaries) are welcome as well.

General information (first three lines)

Sex/age

Feather-length, weight, fat score: this is optional on ringing stations where this information is recorded in the ringing lists.

Skull pneumatization: This is important additional information. Give either score or tick the appropriate box.

6.6.4 Moult Data

Always fill in one side (wing and tail) completely. If time allows, complete both sides, especially if they are different or deviate from "normal" moult patterns. If you hold the bird with the head towards you, you might turn the moult card.

Write codes into the white feather boxes. Use horizontal lines to indicate that the same code applies for a series of feathers.

Give the approximate percentage of old, new and growing feathers for body feathers, marginal ("lesser") and median coverts. If body feathers are composed of three generations, the additional column may be used (explanations in the comment section).

Codes: The aim is to assign each feather to the moult when it has been grown. Codes 0 - 5 are the same as those used in the BTO moult card.

0 – old
1 – feather missing or pin
2 – feather just emerging from sheath or up to 1/3 grown
3 – feather between 1/3 and 2/3 grown
4 – feather more than 2/3 grown, but still not full grown or with trace of sheath remaining at base
5 – renewed in summer/autumn in the breeding range (postbreeding/postjuvenile moult)
6 – renewed after autumn migration during (late autumn) winter/spring ("prebreeding" moult)
7 – In winter/spring: older than 6, either 0 or 5. This code may be used in winter/spring for feathers that appear much older than 6, but for which it is uncertain whether they have been acquired during the postjuvenile/postbreeding moult or earlier; In summer/autumn: older than 0. This code may be used in late

summer/autumn for feathers, which have been retained during the previous prebreeding moult (e.g. adult Spotted Flycatcher, Golden Oriole)

8 – older than 6, either 5 or "early 6". This code may be used in winter/spring for feathers which appear somewhat older than 6, but for which it is uncertain whether they have been acquired during the postbreeding/postjuvenile moult before autumn migration or during an early "prebreeding" moult in late autumn/winter

9 – impossible to assign".

6.7 Field Ringing/Data-Collecting Form

Field data collecting forms used in the station work are specially designed sheets bound into field-books, which prevent accidental loss or damage and allow easy handling of the contents, both in the field and for data input.

Note: It has been suggested that field data can be directly entered into laptop computers, but such a procedure is **extremely unsafe** because of high vulnerability to typing errors – everybody, even accidental helpers, is capable of writing numbers and text to the form in a correct way, while only well trained people can type quickly, and accurately, on a keyboard.

The Network basic data form contains a space for main ringing data, a standard set of measurements/scores and additional data fields for optional data collected at the particular station.

Filling-up the ringing field-book:

1. The cover page (Figure 6.24) -
   - Station symbol
   - Year
   - Season (spring - autumn)
   - Running number of the field-book

   Note: there could be up to three field-books used simultaneously, numbered separately: two for the most commonly used ring sizes (types – see below) and the third – for all other ring types. One field-book, for all used rings, can be used as well, but this would be more time consuming to input into a computer.
   - Date and hour of the first and last item noted
   - Ring type and ring numbers (from – to) included into this field-book
   - A box filled-up when data are already entered into a database (signed by the person who entered the data).

2. The front page (Figure 6.25) has a space for listing the ringers, with their codes written to the form and their periods of ringing. It is highly recommended to have only one responsible ringer at a time.

   Special events affecting ringing,e.g. extreme weather conditions, loss of nets, low number of staff, etc., should be noted in the lower table on this page.

SE EUROPEAN BIRD MIGRATION NETWORK

Station	Year	Season	No.
RINGING			
From:		To:	
Date		Date	
Hour		Hour	
Ring type	Ring numbers		
	–		
	–		
	–		
	–		
	–		
	–		
	Data input:		

Figure 6.24: Bird ringing notebook – cover page.

CODE	Name	From (date, hour):	To (date, hour):

SPECIAL EVENTS affecting ringing (extremal weather conditions, loss of nets etc.)

Date	Details

Figure 6.25: Bird ringing notebook – front page.

3. The main data sheet is spread into two neighbouring pages (Figure 6.26). Each individual set of data is written on one row, divided into columns containing the data in a sequence. The division is adjusted to the sequence dictated by the ringer (see *Laboratory Working Routine* - p. 118).

The same sheet is used for both ringings and retraps/controls – this simplifies input of the data and saves time.

Each column is characterized by one of three special proprieties, symbolized by a special character below the column head:

| – vertical strokes or lines are allowed when the content in subsequent rows is repeated. This speeds up the filling of columns containing data frequently repeated for many individuals: hour, net no., species code and sex,

o – vertical strokes are not needed in the column: Date is default, the same as at the beginning of the page unless part of a page is demarcated by a horizontal line and a new date is entered. Ring type and ring series are by default the same unless specified by new input. Ring number must be filled for all individuals. Status is empty by default for all newly ringed birds. Ringer code is assumed to be the same for all records on the page unless the ringers change.

x – vertical strokes or lines are not allowed in the column – in some cases (Age, Fat) a stroke could be misread later as letter "I" or number "1" or when repetition of the same value is rather rare (Wing, Tail, Weight).

Date	Hour	Ring (o)			Stat	Net	Species code	Sex	Age	Fat	Special data			
o	|	type	serie...	no.	o	|	|	|	x	x				

Wing-formula (x)									Wing	Tail	Weight	Rngr	Comments
(tip)						(8)		(10)	x	x	x	o	o

Figure 6.26: Bird ringing notebook – left and right pages.

Subsequent columns contain:

Date – formats allowed: e.g. 1.9, 1.09, 01.09 (Sept. 1st); it must be written at the beginning of every page and when the next date starts.

Hour – formats allowed: e.g. 6 or 06 (for a nets control at 6.00); full hours only.

Ring type – one or two letters (or up to two digits when ring type is described by first number digits); when only one type of rings is noted within a sheet, only the first box on the sheet should be filled. When more types are present, write the type when changed; at retraps and controls, write the type in each case.

Series – newly ringed bird: all digits of a number except last two; if only one type of rings – write only once at the beginning of a page; retraps and own controls: write full ring number here.

No. – newly ringed bird: last two digits of the ring number; retraps and own controls: leave empty.

Stat. (status) – at Operation Baltic and SEEN stations there are used two groups of codes, while some other status code systems could be used.

1. Birds caught within normal routine work

– Leave empty for newly ringed bird,

R – Retrap (a bird ringed within the same season),

C – Control (a bird ringed elsewhere in the country or at the site during previous seasons),

V – Foreign bird control,

N – Released without ring,

D – Bird dead during catching/ringing, this code letter could be added to any other code, forming two-letter code, e.g. RD – retrap dead.

There are a few sporadically used codes for the bird status:

Z – Ring changed, A – ring added are used exceptionally when existing ring is damaged or heavily worn (if it is possible to remove it without risk of injuring the bird: change the ring, if not: add another ring on the second leg). Never put the second ring on the same leg.

2. Birds obtained outside of the standard catching procedure (found, caught accidently, brought by visitors, etc.)

F- All such special cases when the bird will be not counted to migration dynamics data,

X – None of the above listed,

D – Code is applicable as a second letter.

Net (optional) – net number (if appropriate) or net symbol (if special net type)

Species code – five or six letter code.

Sex – M (male), F (female) or zoological signs (♂, ♀); leave empty when not known.

Age – J (juvenile), I (immature), A (adult), N (not defined) (see p. 87)

Fat – fat score 0 – 8 (see p. 95)

Special data – optional, according to the station needs.

Wing-formula – (tip) – numbers of primaries (ascendant) being the wing-tip, e.g. 3, 34 (3=4), 35 (3=4=5) etc. Distances of subsequent primary tips in relation to the wing tip;

For outer (distal) primary measurements, add "0" (zero) at the beginning;

When distal and proximal primaries are equal, write the same number in two subsequent boxes;

(8) Last box when standard (to the eight primary) method used.

Wing, Tail – write in full millimetres.

Weight – formats allowed: e.g. 16 or 16.0 depending on accuracy of measurement (to the nearest gram or to the nearest 0.1 g respectively).

Rngr – ringer code (once per a sheet when no changes).

Comments – plain text comments or the station free use.

7 Passerine Station Laboratory Working Routine

An optimal laboratory routine is essential for the collection of high quality (and large amounts of) data without danger to the birds. This routine should be as parsimonious and as effective as possible. On one hand, it should be flexible, and on the other hand, separate operations should be strictly standardized in order to be compatible.

A few routine levels will be described:

1. *Normal routine* where a standard set of data is collected,
2. *Extended routine* where all (planned) additional data are collected,
3. *Reduced data collecting routine* as a part of the "*alarm routine*" when too many birds are caught to perform standard working procedures.

It must be stressed here that "normal" and "extended" routines may be differentiated according to the station preferences, based on its scientific scope of work and/or concentration on different groups of birds.

Depending on the number of birds waiting for treatment and the existing routine, the work performed at the laboratory may be organized according to one of two guidelines: (1) if the number of birds is low or moderate, the working team is made up of two people, and (2) when birds are numerous, the team is made up of three people. The experience of the people involved and their training in the work as a team collectively decides the meaning of "numerous".

Note that it is very inefficient to have the same person ringing and at the same time entering data; this situation should be avoided as much as possible. It is very time-consuming and can lead to errors.

Generally, the laboratory routines are based on strict attendance to the rules as earlier presented, particularly those in the *Netting* and *How to Arrange Trapping with Heligoland Traps* sections. As a short reminder: the birds transported to the laboratory are hung in bags on a row of hooks at the edge of the laboratory table (Figure 3.14) and they are sorted by species (only one species per bag and bags with the same species in a row) and by ring sizes (all species ringed with the same ring type should neighbour at hooks). The sequence of ring types should be permanently fixed in order to obtain the same working procedure, e.g. the smallest type always to the left of the row and subsequent sizes following to the right. The position of the bags is decided by the present working team. The seats of people in the working procedure must be placed side by side, with the ringer seat to the left (right-handed persons assumed). The seat for a third person in the three-person working group should be located at the other side of the table so that the writer is able to hear dictation from both working persons. The most useful seat for the ringer is a soft, comfortable armchair, for you sometimes spend a few hours ringing when the birds rush. The seat should be of such height that the ringer's thighs are situated horizontally when he sits with knees close together and feet a little bit apart. This position will allow the ringer to put his ruler on his lap while not in use. Seats for the other people should be more elevated, in order to make writing on the field-form lying on the table comfortable enough. Sometimes, in the two person procedure, it is preferable to have the field-form resting on a hard surface situated on the lap of the writer, while he sits in exactly the same position as the ringer – this way, it is more comfortable to pass the birds to the ringer.

7.1 Normal Routine

7.1.1 Two-Person Procedure

Out of the two people working in the team, one is (as a basic rule – see below) designated as the ringer. The ringer rings and takes all measurements, while the second person acts as the writer, noting dictated data in the field-form. In order to simplify, record the ringing/measuring procedure in a strictly defined order, fixed to the sequence of columns in the field-form. Strict standardization of the working procedure is very useful when many people are taking part in laboratory work, changing their roles while bird processing, or alternating between bird stations working within the network.

In normal procedure, the ringer removes the bird from the bag (holding the bird in the standard manner) and starts with ringing. It is a good custom to start a new hour ringing with the same species which was the last input of the particular field-form, e.g. if the last bird ringed was a Robin – start with the robins, if there are any. This saves time when entering data into a computer file. If this rule causes any difficulties (we do not know in which bag the Robin is) – disregard it.

Standard dictation goes as follows:

Hour – this is dictated only when birds from more than one control are waiting. Otherwise, *Date* and *Hour* are written by the writer without dictation and it is his/her responsibility to make a correct input.

Ring number and **Status** –

1. For a new bird, not yet ringed, the ringer dictates ring type and the two last digits of the ring number; if the ring is the first one on the ringing-sheet, the writer calls for a full ring number and the ringer provides it; after that, the writer must check whether the number given is a subsequent ring number. If not, the writer must stop the procedure and the problem must be solved (lack of a ring? wrong sequence of rings? false ring type? new series? retrap or control?). The writer is responsible for noting the correct ring number. The first part of the ring number (ring number without last two digits) is written into the column "*Series...*", the last two digits into column "*no.*". Subsequent numbers in row are noted as only two last digits in the column "*no.*"

Note: the ringed individual is then measured according to the established standard (see below).

2. A re-trapped bird is reported by the ringer first as "retrap", followed by the full ring number with heading letters (ring type). The writer notes the type of ring in the column "*Type*" and full number into the column "*Series....* ". Then, he writes "*R*" in column "*Status*".

Note: re-traps are usually not measured, but fat score and weight are noted. However, measuring re-traps could be useful for studies on measurement calibration of ringers working at the station and ascertaining the reproducibility of measurements.

3. Control: a bird ringed elsewhere with a ring issued by the same ringing centre or ringed at the station in previous seasons. The ringer reports it first as "control" and dictates full number; then he asks the writer to read the recorded ring number and compares the recorded number with the ring. For controls, as for re-traps, full ring number is written in the column "*Series...*". The status of such control is noted as "*C*".

Note: Controls are measured according to the newly ringed birds standard.

4. Foreign control: the ring number is noted as usual but re-dictation is obligatory; the ring number last column should be filled with sign "–". Status is noted as "*V*". In "*Comments*" it must be written as the full inscription on the ring (double reading should be applied).

Note: Foreign controls are measured according to the newly ringed birds standard.

Species name: is coded as explained earlier. The ringer must dictate the code – not the bird name, unless the writer is a qualified ringer. If same as the previous individual, the bird name may be left out and the writer fills the name position with a vertical stroke. Note, however, that this is a slightly dangerous custom. The writer should be cautious and he should check with the ringer. Generally, it is allowed only when the writer is well trained.

Sex and **Age**: dictate codes unless the writer is trained; sex should be dictated with words when scientific symbols are used in noting. Note: sex and age notations can be fixed after fat scoring (the ringer is able to see some sexing/ageing characteristics, e.g. sex of the Great Tit, when studying fat deposits). It is possible just before *Wing-formul*a too – in many cases, opening the wing is needed both for age discrimination (e.g. contrast within greater coverts), and for starting wing-formula measurement.

Fat scoring

Wing-formula

Wing-length

Tail-length

Weight: these are dictated according to rules specified in standard descriptions of methods. Such a sequence is used because a sequence of measurements, done with the same ruler, will save time. It was carefully optimised from the ergonomic point of view, so the bird and the ruler moves are limited. This is very important when the bird count gets high!

Ringer: ringer's code is written at the beginning of every sheet (without being dictated).

This basic two-person procedure can be modified according to number of birds and experience of the writer:

(a) The writer can remove the bird from the bag or box and pass it to the ringer, or, if very well trained – ring it and pass it to the ringer, who becomes the measurer. This option requires remembering of some data and sometimes can lead to errors;

(b) The writer can be the person who reads the result from weighing.

Applying a three-person and non-standard version (a) of the two-person procedure requires quick and safe passing of the bird from hand to hand. Passing of the bird when both people hold it by the standard holding method is quick, and birds seldom escape (Figure 7.1): the bird holder takes the bird's bill with the left hand and gently pulls it. The bird's neck becomes longer, and simultaneously, the right hand turns the bird's body slightly and holds it with the tips of four fingers; the bird receiver creates a fissure between his index and middle fingers of his right hand and directs it to the bird holder. The bird's neck is placed between two fingers of the receiver. The receiver need not look at the passed bird and he closes fingers when he feels the bird's neck between them. The passing procedure should be trained in advance before there is need for it during a rush of birds.

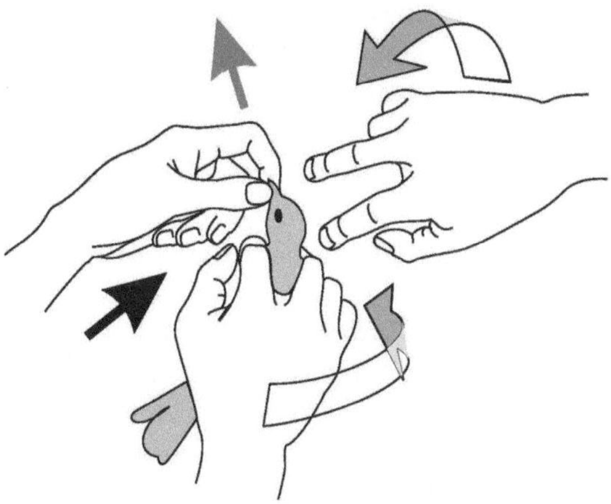

Figure 7.1: How to pass the bird quickly and safely.

7.1.2 Three-Person Procedure

The number of birds that can be processed by a two-person team is limited, so if many birds are caught, a three-person team is even more efficient. One person is the ringer, the second is the measurer and the third is the writer. The course of the procedure is still the same, but two working persons process two birds simultaneously; the ringer removes the bird from the storing device (bag or box), rings it and performs sexing and ageing. The next person does all measurements. Both working persons: the ringer and the measurer, dictate data one at a time, not simultaneously. This is the point at which it is crucial to be well synchronized! The writer must record the data, correctly filling two rows in the field-sheet. The best possible synchronization is reached by dictating the next bird during the weighing process, which is the slowest element in the procedure. The writer's role is difficult but still possible to perform when he is well trained. It is important to have silence in the laboratory (if silence is a possible state when a hundred tits are waiting for ringing!). In reality, silence means "no one talks". A well-trained team is able to process the bird, on average, within 20-25 seconds, taking all standard measurements using recommended methods and strictly following recommended routines.

7.2 Extended Routines

Contrary to normal procedures, where a high level of standardization must be maintained, there are possible arrangements for different extended routines. They depend on the set of data collected and preferences given to the studies being performed. There is more freedom to negotiate these choices when the number of birds is limited and there is enough time to work more slowly. Most problems arise when the number of birds rises and there is competition for time between different studies being performed. This will be discussed in the *Alarm Routine* section. In all cases, however, the routine applied should be defined in relation to the standard working routine.

When there is little additional data collected, additions may easily be incorporated into the normal routine. Additional measurements/scores may be taken after sex/age discrimination, or fat scoring if applicable, and wing-formula measurements. In the field-form, there are optional fields located for additional data. Another reason for making additional measurements all together is that they are usually done with special tools, not a standard ruler. Sometimes, they can be done towards the end of a standard procedure. This is usually the case when they are treated as data of secondary importance, so they may be abandoned with little grief.

When non-standard data collection is more time-consuming, and the study is important, it is advisable to do additional data collection after the standard procedure has been performed on all birds caught. Alternatively, a separate person or team can collect the data. This rule should be applied particularly to moult studies, examination for parasites, blood-sampling, etc. Orientation tests can be done parallel to the ringing, since one specified person usually works with tested birds. He may start after the first individual of the species studied has undergone the normal procedure. Individuals selected for special treatment should be put into unique bags and then be hung separately from other birds.

7.3 Alarm Routine

Occasionally, large numbers of birds can be caught at a station, e.g. at the Operation Baltic station, Mierzeja Wiślana, on the best day 3300 birds were caught (more than one thousand are regular there and elsewhere) while more than 800 could be caught in a single net check. This possibility must be taken into consideration before it occurs. For a stressful situation like this, it is important that the chief ringer, as well as all helpers, are psychologically prepared. Since migration peaks are always sudden, people that are not prepared for a sudden rush of birds frequently lose their sensibility and may do totally irrational things.

First of all, constant maintenance of the rules of normal procedure (described in earlier chapters) is recommended. All elements stressed there synergize with successful solutions of problems caused by an extreme bird rush. Here, the most important advice is summarised:

1. The net round should be as simple as possible and cleared so that workers are able to run along it without colliding with twigs and strings nor stumbling on laying branches, stones, etc.; passages under the nets and strings should be avoided.
2. The nets should be made of a kind of material that allows quick removal of birds. Any nets with very thin thread should be taken out of use when mass trapping is expected; specially designed nets which must be closed or slipped along poles when the birds are removed should not be used at all (or opened only when one is sure that there is no rush of birds). If the rush comes suddenly, such nets should be immediately closed.

   The nets must be clean of leaves, twigs, etc., and not caught on trees and bushes.
3. The equipment should always be ready for use. The number of necessary dry bags, boxes or baskets should be adequate; it is a good custom to have a special reserve of fresh (never used before) bags ready for a special situation like a rush of birds, particularly in wet weather. Sometimes, if many birds are caught on evening controls, then, a good source of light should be available.
4. There should always be enough rings of all sizes: opened and ready to use, particularly of the ring sizes most commonly used; it is better to have a surplus of a thousand rings than a hundred too few. Having to open rings when the stock is exhausted can be time consuming and problematic.
5. The staff must be trained in the correct removal of birds from nets, in selection by species, and in hanging bird bags in proper places in the laboratory. Furthermore, they should be trained in how to pass birds from hand to hand and how to register data in the field-forms on days with a limited supply of birds. Individual aversions to drill and dull, standardized work should be overcome, people must be carefully trained and motivated to useful routines so that they can "save the birds from death" when a rush occurs. NOTE: the "real" rush (with several hundred birds at

one single control) is hardly imaginable to people who started their practice at a station where twenty birds are caught per day!

6. At the outset of the control walk, particularly the first one in the morning, always bring many more bags than are probably needed. On a peak day, the actual need may be ten times (or more than that) higher than on a normal day. Lack of bags may seriously disturb the rhythm of controls and has been the cause of birds' death, when too many birds are put together because the helper doesn't want to return for more bags, in more than one case. Therefore, it is a good custom to assess the number of birds in the first nets and estimate whether there could be a coming rush already at the outset of the control walk. If there seems to be a need for more bags, return quickly to the station and warn the chief-ringer about such a prospect; sometimes, the expectation may turn out to be unfounded, in other cases it will save birds' lives and the ringer from the stress resulting from the collapse of routines (and a load of responsibility for the birds). Also, keep in mind that the staff will be grateful for smooth work and smooth routines; at the beginning of a day, when the ringer gets notice of an approaching rush, he should immediately wake up all personnel and order who will be a writer and who will go where.
7. Remove the birds by species from the nets when there are many birds of two or more species. The bag should not be closed after each bird; see *How to free a bird...* – p. 45. Strictly follow the rules, restricting the number of birds in a bag and their transportation. The decision of whether to put the birds into boxes or baskets is made by the chief-ringer. Do not mix birds from two controls; it must be absolutely clear which birds were collected simultaneously.
8. In most cases, the net control is done by one person or by two attending training together, or simply for social reasons. When a rush occurs there are two possibilities when two persons work on the same control path:
   (1) Many birds are known to be waiting for removal and the staff is numerous enough to allow two persons to work together on one control path. They walk together following the normal course of the control (a fixed direction of walk); remove different species when working together at one net or work on two sides of the net when birds have been caught from both sides,
   (2) One single person goes for the control walk and does not return within the expected time. In such cases, the second person sent by the chief-ringer must go in the opposite direction, and when he runs across the first one the two together should return to the laboratory without removing new birds from the nets that are passed for the second time.
9. Keep a time schedule of the control walks! Keeping the time schedule during a peak of catching means being no more than fifteen minutes late. Remember that birds staying in the nets for a longer time get more entangled and their removal takes longer.

10. Work with the three-person or at least the two-person procedure at the laboratory; one-man-work is highly inefficient. There may be a gain of momentum if the ringer assists in removing birds at the first control walk and then returns with a helper to work in the laboratory as a two-person team. Since there are enough birds for continuous work, and handling speed is a key factor on peak days, the ringer must then stick to the laboratory with one or two helpers, according to the procedure, because if the whole staff concentrates exclusively on removing birds, not before long bags and other storing devices will be full of birds waiting for treatment. People working in the laboratory should be asked for silence. The working team should receive only necessary directions and information. Any additional voices disturb the rhythm of dictation and can lead to repetition of measurements and errors in writing.

There is no single formula for winning the battle with hordes of birds in such a way that all requirements of this strategic game will be fulfilled. The birds fly to their migration goal ringed and measured, while you and your staff will be still alive and satisfied because of the high quality of the work done.

Finally: all people must be psychologically prepared to make a maximum effort at any position: as ringer, writer or helper. The most important thing when a rush occurs is that the chief-ringer does not panic; this usually leads to unwise decisions, resulting in avoidable deaths of birds or at least in unnecessary losses of data, since data from peak days could be of great scientific value.

There are a few general observations, which could be helpful when it comes to evaluating a "rush" situation:

(1) In practice, a really huge rush of birds seldom lasts more than three to five hours, so for a moment you may be close to making desperate decisions. This stage is usually reached during the third hour of the rush. You are also close to a report from the helpers „we have ten birds from the last control walk". The timing of peaks differs a little and is both species and site dependent, e.g. at Mierzeja Wiślana, Poland: thrushes: only at first control walk, the Robin: the first two walks, tits: mostly the three to four first hours (sometimes later in the day, but for a shorter time), the Goldcrest: three to four hours, but starting from the second control walk etc. But on a grey and misty October day, goldcrests may also move about till dusk; this means 12 hours of uninterrupted ringing by exhausted staff! The possibility of reversed migration in the afternoon should always be kept in the corner of the chief ringer's eye. An outbreak of starving siskins will last from dawn till dusk, and spring arrivals of chaffinches, robins and goldcrests on islands in the Baltic and Kattegat may last well into the afternoon; a number of ringing catastrophes are known to have occurred under such conditions. It is recommended to have a look at earlier catching files of the station in order to learn the patterns of different species.

(2) Dead birds are inevitably connected with the numbers caught during catch. When more birds are caught, the theory of probability tells us that there will be more dead birds. In addition, the probability for unavoidable deaths, due to predation, strangulation, and exhaustion, is higher on peak days than on quiet days. When few individuals are caught; the total catching mortality is a weighed as the sum of these probabilities and it never equals zero. On a peak day, accidental losses, like a couple of dead birds in a bag, caused by e.g. instinctive stronger hold of a bag full of birds falling down from the bag hanger, are much more probable. Such singular losses, however, are a far ways away from a real ringing catastrophe.

The chief-ringer is the only decision-maker at the station unless there is another well-qualified ringer to whom the chief-ringer could pass responsibility to decide on duties of the staff members. The decision-maker must be well informed about what is going on in the field: how many birds there are from the current control and if the birds are very active, for instance. Apart from decisions concerning the organization of work, the chief-ringer must make some other key decisions alone, and these depend on his appraisal of the staff's abilities to cope with the expected number of birds, taking into consideration the expected time distribution of other potentially occurring species (see above). A few standard emergency decisions will be given here in order of importance:

1. close special nets that are time-consuming when active. If these nets are not designed for a special study, the decision should be automatic when a rush of birds is observed,
2. stop any additional data collection. This should be done as soon as it is obvious that the day is a peak day, unless the data have a very high priority within the station programme. In such a case the following decision (3) could come first,
3. Stop taking standard measurements. It is very important and must be emphasized: stop all standard measurements at once, not in a few steps; only ringing and sexing/ageing should be continued; this is a key decision for the data collection and it should be undertaken in a situation when there is real danger to the birds (but not because we are hungry and tired!): (1) the birds become weak because of poor physiological condition, low fat reserves caused by a long flight and/or bad weather. Some exhausted birds are always observed during intensive migration, and they are selected both by catching stress and natural migration risk. A really dangerous situation occurs when ringed and released birds do not fly away but stay around the ringer, most of them after a short rest go farther, but some die. The key species is always the species in the worst condition; (2) The ringer estimates that he is not able to ring all waiting birds within a reasonable time, even if the birds seem to be in sufficiently good condition (Table 7.1); (3) The birds are wet and it is dangerous to have them stored in bags, and (4) There is lack of bags and storing devices which may cause disturbance in the rhythm of control walks.

4. When the weather is favourable, and there is a group of nets with much lower catching ability at the end of the control path, it is possible to have them checked only every two hours. This is an exception and the decision should be based on a good knowledge of local catching distribution. The rush in itself may mean new conditions of these nets!
5. The most difficult decision is to close the nets; this always means an interruption of the seasonal dynamics of the station and a vacant space in its monitoring data; in addition, the proper closing of nets takes time that could be spent on removing birds from other nets. These birds will have to wait longer and hence become more entangled.

Every emergency decision should be cancelled as soon as the chief-ringer estimates that the situation is no longer dangerous: the number of birds waiting is low enough, the individuals are in good condition and the rush is over. It must be remembered that *when the standard set of measurements is started anew, the set should be complete.*

When the rush is over, not earlier, the staff may quietly have its lunch (for breakfast it is too late), clean the laboratory (there is usually a mess of lost feathers, excrement, dirty bags and boxes) and start to prepare rings for the next day (if you need to do so), peak days often come in a sequence, e.g. once, at Mierzeja Wiślana, Poland during 21 days there was an average of 1000 catches daily. During such circumstances it may be worthwhile to wake up people earlier than normal and let them have breakfast prior to the first control walk.

Table 7.1: Time limits for storing caught birds.

Most common fat score	*T0*	*T1*	*T2+*
Time limit*	2 h	3 h	4 h

*Counting from the nominal hour of the control of nets (e.g. birds from the control at 6.00 and scored as *T1* should be free till 9.00)

PART II: **The Wader Station**

At the wader ringing station, birds are usually caught in walk-in traps or in mist-nets. In addition, other catching techniques might be used as well. The work with traps and nets is completely different from one another. Mist-nets may be used in parallel to walk-in traps, particularly after dusk when waders usually do not walk into the traps. However, in a longer period, it needs two ringing teams at the ringing sites.

8 Wader Catching Techniques

8.1 Walk-in Traps

Walk-in traps are selective catching devices, and their use produces some biases. They are less effective for long-legged species, however, models with larger dimensions were quite effective in catching large waders. In addition, waders with a visual foraging technique (e.g. *Charadriidae*) are not as easy to catch as tactile feeders. Thus, the structure of species of trapped birds does not reflect that obtained from counts. Moreover, there is some evidence that birds avoid the sites of initial capture. Thus, results of studies which involve multiple trapping of birds should be viewed with caution (Muraoka & Wichmann, 2007). Walk-in traps are very convenient for use and are safer for waders than mist-nets. Birds inside the trap are often unaware of being caught and continue feeding or take a nap until people approach. Moreover, catching in walk-in traps is almost independent of weather conditions, except those influencing the water level. It is worth noticing, that this kind of trap could be used also for catching wagtails, pipits, rallids and small dabbling ducks. Walk-in traps could be placed in different habitats, e.g. on sandy seashore, at small, shallow muddy bays, sewage farms, wet meadows, but it is inconvenient to use them in areas with regular tides. They differ in shape, dimensions, localization of capture chamber, types of entrance and material used for their construction (wire netting or thick fish netting). A variety of different types is shown in the book of Bub (1991). In Poland, walk-in traps were used since 1960 (Figures 8.1-8.2). Generally, three types of walk-in trap constructions may be distinguished (Figure 8.3). The first one has the capture chamber located at one side and is described as safe, limiting the mortality of trapped birds due to rapid changes in water level. The second one has two capture chambers, while the third one has one capture chamber placed in the middle. Traps showed in Figure 8.3 could be built as lightweight constructions, which make it possible to carry them by one person. Usually they are made of wire frames and a cover of thick fishing net (thread of no less than 1 mm thick, mesh 16 - 19 mm). Fish netting is less durable than wire netting with a protective zinc surface, but the latter causes more injuries and plumage damage in trapped waders, particularly, to snipes when they run their heads against the roof while fluttering inside the trap. It is convenient to have some standard sizes of frames, which makes maintenance of traps easier. For example, the walk-in trap can be built from frames: 80 × 40 cm (roof and gable walls), 35 × 40 cm and 50 × 40 cm (flaps and walls attached to funnels), 150 × 40 cm (guiding fences). Frames are joined with pieces of wire, but different fixing methods could be applied (e.g. cable ties). It is important to have at least two fixing points between the two frames, which increases the strength and durability of the construction and protects it from damage when moved. Both models may be produced in folded and unfolded versions. In unfolded versions, less wire is needed, but this model is less convenient

Figure 8.1-1: The first wader trap used in Poland. Mouth of Vistula 1960. Photo P. Busse.

Figure 8.1-2: Wader traps in Poland were quickly developed. Mouth of Vistula 1962. Photo P. Busse.

Figure 8.2-1: Traps used currently. Mouth of Vistula. Photo W. Meissner.

Figure 8.2-2: Traps used currently. Mouth of Vistula. Photo W. Meissner.

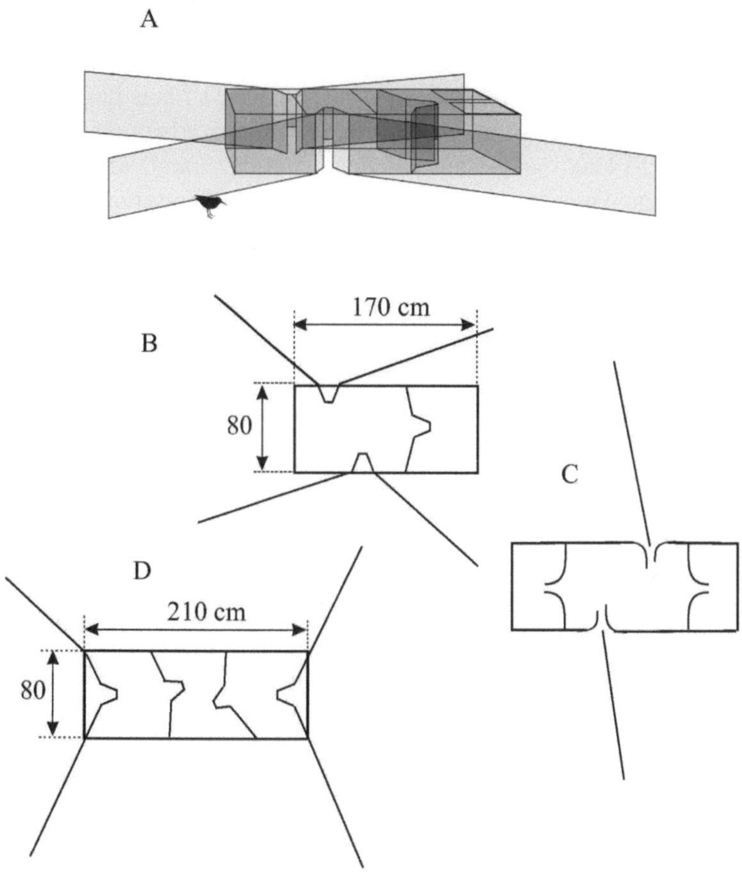

Figure 8.3: Different types of walk-in traps for catching waders, A. with central capture chamber, B. with capture chamber at one side, C. with two capture chambers and curved-wall entrance, D. overall view of walk-in trap of recommended type. Measurements are in centimetres.

for transportion and storage. Moreover, when the net is damaged (at some point), it is easier and quicker to exchange one frame from the folded version trap than to take the whole trap from the catching place to be repaired. All frames should be made of stainless wire, preferably zinc-plated. It makes them last longer, especially in marine habitats. Unprotected wire of 5 mm diameter may rust completely after 4-5 years of use.

The optimal total height of a trap is about 40 cm. Such a trap will catch a wide spectrum of species (up to Oystercatcher size). Higher (e.g. 50 cm high) and larger traps seem to frighten off smaller wader species, but they are quite effective in trapping the larger ones.

The form of the entrance is essential for optimal function (Figure 8.4). According to some opinions, a funnel-shaped entrance with proper "depth" is better than the „curved-wall" type (Figure 8.3C) (Meissner, 1998), while other authors had opposite opinion (Lessels & Leslie, 1977). In the case of the narrow space between the walls of the entrance (e.g. 2-3 cm), escapes are less likely, but only small wader species could be caught. The main problem with "funnels" concerns their position. Entrances should not be placed in front of one another. In this case, the first entrance will lead the bird directly to the "exit" so that a bird, once caught, easily escapes from the trap. The narrower the funnels, the less chance for trapped bird to find an exit. Dimensions of the funnels presented in Figure 8.4 allow catching waders up to the size of an Oystercatcher, but small species have greater chance to find an exit. The funnel, which leads to the capture chamber, can be narrower as birds force their way in.

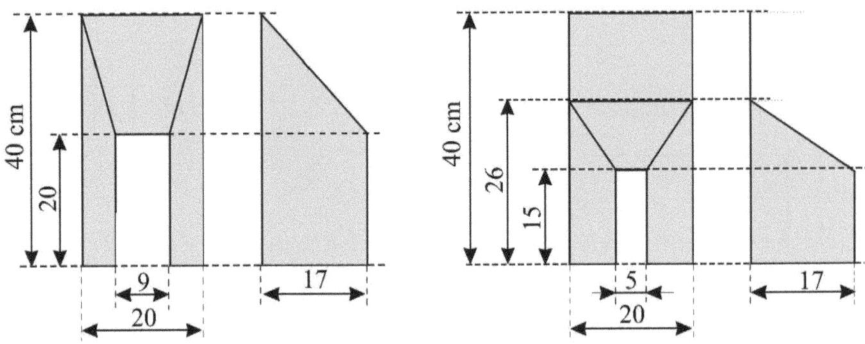

Figure 8.4: Funnels for walk-in traps. Measurements are in centimetres.

Another important part of the trap is a "guiding fence" that leads the foraging birds to the funnel. It is convenient to have fences fixed to the trap, because it makes redistribution of the traps easier and quicker. Other options are to have fences with additional "legs" to put into the ground (Figure 8.5). The fence is the most susceptible part of the trap, particularly in marine habitats, so it may be useful to have spare fences stored at the ringing station.

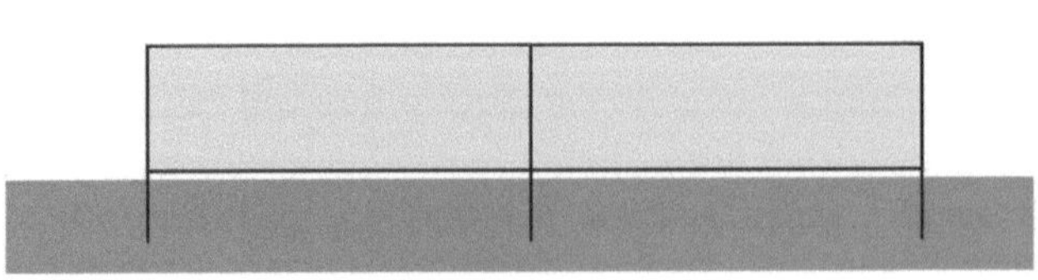

Figure 8.5: Guiding fence for walk-in traps.

Flaps should be fixed to only one of the neighbouring frames (to be opened quickly). It is necessary to close flaps with a hook because trapped birds may open it when fluttering against the roof inside the capture chamber.

The other type of walk-in trap (the so called "tent-like" type) is quite effective, and when folded, it could be carried even in one's pocket. Traps of this kind are made of fishing-net and thick rope, and frequently erected along the edges of small pools (Figure 8.6). When making this type of trap, it is necessary to be very careful when cutting the netting. Instead of dimensions in cm, the number of meshes should be counted to have all parts fitted to each other. The same concerns fences. The tent-like type may be inconvenient in cases when the trap must be moved often from one place to another. Therefore, it may be used in places, for example, with a stable water level.

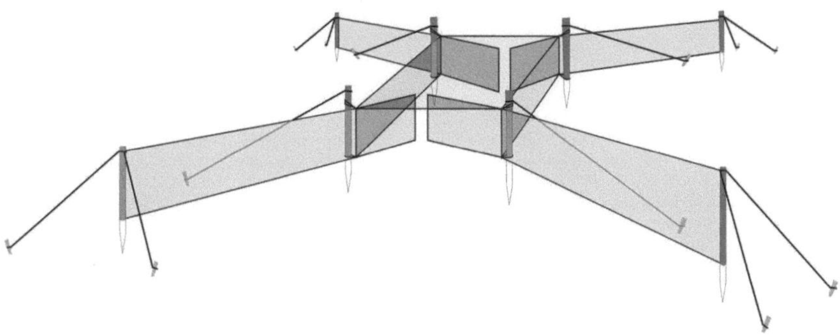

Figure 8.6: Soft netting ("tent-like") walk-in trap.

8.1.1 Arrangement of the Catching Area

The method of setting traps depends on many elements regarding the habitats in which they are used. First of all, it depends on the presence of a clear boundary between water and land. The traps should be set only where birds forage, for resting-places are no good for catching. At the shore of a pond, a lake, or in the open sea, traps should not be set in direct contact with the water but rather placed a bit "inland", while the fence prevents birds from passing them on the water side. If the shore bank is wide, the setting of other traps (or additional fences on the landside) will remarkably increase catching efficiency (Figure 8.7). In places where there is no clear demarcation between the body of water and more or less dry land (mud, wet meadow), traps not linked with each other will not be very effective. In such cases, a line of several traps with extra-long fences will give the best results. A V-shape arrangement (like on a water pond bank) as well as a single line of fences starting from the inside of the entrance may be used (Figure 8.7). Walk-in traps could be used without guided fences as well. However, in most situations, their catching effectiveness is lower. It is recommended to observe first where waders feed and then put traps there. In flat, open areas with no clear leading lines, waders foraged into the wind. It is also better to put walk-in traps with fences perpendicularly to the wind. At new locations, where there is no knowledge where waders feed, the best idea is to put traps in place with many footprints.

It is important for the safety of trapped birds to have the capture chamber set in a dry place, or else some sand or cutgrass may be put in it. In this way, birds trapped will be dry even when the traps are placed in mud or on a wet meadow. Lightweight walk-in traps may be settled on floating beds of seaweed or other water plants as well, but there is a risk that they will sink within a few hours. In such cases, wood poles should be placed under the trap along or perpendicular to its longer walls.

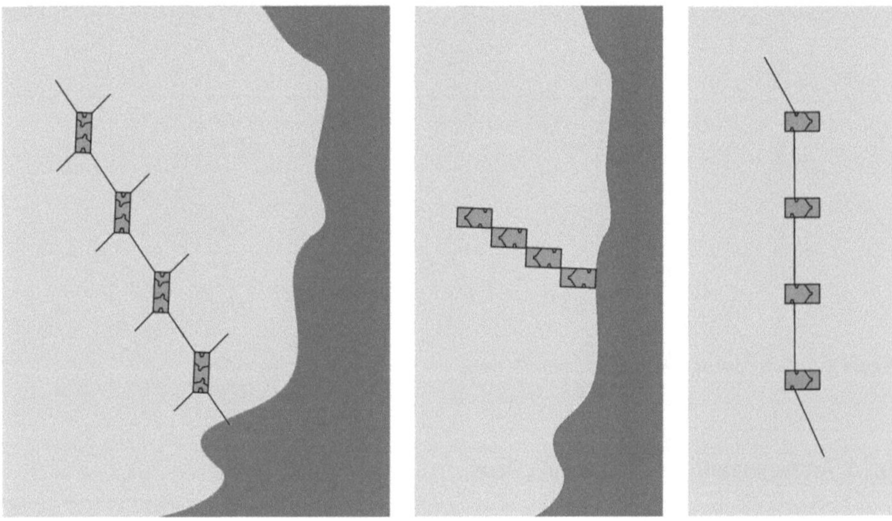

Figure 8.7: Arrangements of walk-in traps at the catching area. Placement of walk-in traps without guided fences was also shown (dark grey– water, light grey - sand or mud).

8.1.2 Maintenance of the Traps

When pieces of water plants or soft mud float in the water, the trap becomes caked with them very quickly. "Dirty" traps are less effective and should be cleaned to make their walls transparent. This is very laborious; sometimes, it takes an hour to clean them. In such a case, it is better to remove such a trap from the catching area. If there are many dirty traps, it is better to limit the catching and have a cleaning session with the cages. This will disturb the rhythm of controls and lower the number of birds caught, but in the long run it can't be avoided, since birds are unwilling to enter traps caked with vegetation or mud.

When controlling the traps, always bring pieces of wire and rope for small reparations and carefully look for damage to the net covering the traps. Even one single broken mesh may serve as an "exit" for small stints and wagtails, particularly, when the hole is situated in the corner of the capture chamber. If only one mesh is broken, it may be quickly repaired in place. Otherwise, the whole trap must be replaced at the next control walk. If there is extensive damage, the trap should be taken to the camp and repaired there. Be careful to close all flaps (covering the hole through which birds are taken) and always keep the capture chamber dry.

In areas where rapid changes of the water level are expected (e.g. rivers, marine environments) traps must be removed when the water level is rising. The number of staff and traps at a particular ringing site should be adjusted to this need. Since there is a constant need to remove, clean and repair traps (particularly in marine environments), 2-3 persons should take part in each control walk so that all these tasks are done quickly and efficiently. Caught birds cannot wait until traps are cleaned, so they have to be taken as soon as possible to the ringing table.

The water level of places where the traps are set should be continuously monitored, and in the case of a rapid increase, one should not wait for the control time to move the traps. It is convenient to fix a suitably marked pole, visible from the camp, so that the water level can be easily read from a distance. Ringing stations situated on riverbanks or in marine areas are well advised to tune in to reports about water levels in the river (or fishermen's weather forecast). This will allow the staff to forecast major changes in water level. If a rise in water levels is expected overnight, it is better to move the traps beforehand to prevent them from being flooded. Watching the behaviour of birds from a distance can assess the effectiveness of the walk-in traps, e.g. to prevent escapes from birds walking through the trap and out of the opposite funnel, adjust the position of traps within the catching area and check if funnels are placed correctly.

8.1.3 Control of the Traps

It is worthwhile to have the traps arranged so that they do not need to be passed twice. This creates unnecessary disturbance to the birds. Additionally, try not to set traps too densely, because this may discourage birds from feeding at that particular section of a beach or riverbank.

Walk-in traps should be controlled every two full hours. Some authors (Lindström *et al.*, 2005) claimed that traps should be emptied every hour or as often as the ringer likes. However, in many stop-over sites, controls made more often result in scaring birds from catching area. In good weather conditions, it was found that traps could be left as long as 4 hours between emptying (Lessels & Leslie, 1977). The first visit should be made about 1 hour after dawn, the last no later than one hour after dusk. A control walk should not last for longer than 30-40 minutes. Exceptions may be situations where it is necessary to move or clean the traps between the controls.

Even then, however, the break between the subsequent controls should not exceed 2 to 2.5 hours and caught birds should be delivered to ringing site as soon as possible.

Waders will enter walk-in traps not only during the daytime. In some areas, (e.g. in late autumn in temperate zone), waders very often feed through overcast and even during dark nights. Trapping under such circumstances often may be more rewarding than trapping in broad daylight. It is very important to check this possibility from time to time. When night foraging occurs, traps must be checked every two hours the same way as in the daytime.

Sometimes, it may be necessary to temporarily cease catching (e.g. during long-lasting heavy rain, at night to avoid mammalian predators). It is enough to close entrances of funnels by moving guiding fences and "stick" them to the trap wall. Moreover, flaps must be open to allow the flight of trapped birds from the capture chamber. Surprisingly, some birds can enter closed walk-in traps by very thin space left between trap wall and the fence.

8.2 Mist-Nets

Mist-nets for capturing waders are effective only in particular conditions. Net visibility is a major problem. Waders most often feed or roost in flat-open sites where there is no background to make mist-nets poorly visible. Thus, only when they are feeding in overgrown areas could mist-netting be effective in daylight. In open areas, dusk or darkness is essential for successful catches. That is why this method of capturing waders is used mainly at night. However, bright moonlight considerably lessens the catching success. Unlike walk-in traps, mist-nets are ineffective in wind. Hence, results of catching with mist nets depend strongly on the occurrence of fine weather conditions. During windy and rainy nights, catching with mist-nets should be abandoned. Surprisingly, in misty weather, when visibility is poor, the number of waders caught decreases because in such weather, waders' movements are reduced. In general, conditions for mist-netting are more often favourable at inland ringing sites than coastal areas (where windy nights occur much more frequently).

A standard wader mist-net has 30 mm mesh, three or four shelf, 110d/2 ply and the length of 18 or 20 m. This kind of net catches most wader species, but also catches ducks, terns, gulls and rallids. Sometimes, shorter mist-nets (12 m long) are used, but waders may pass around mist-nets rather than go between poles (due to the smaller distance between poles). Smaller species are often badly tangled in 30 mm mesh, so for catching stints, phalaropes and sandpipers (up to female Ruff size), the 22-25 mm mesh seems to be better; however, larger species usually escape from mist-net shelves.

Mist-nets for wader catching should be erected with maximum tension (much more than in the case of passerine mist-nets). When heavy birds, like curlews or oystercatchers, get entangled in the net over water, the net should not sag excessively. It is advisable to check the net tension before catching by putting into the shelf an object with a similar weight to that of the heaviest species is expected to be caught and see the extent of sag. Another idea is to put bi-forked sticks under the lowest shelf of the net every 1-1.5 m. This protects the shelf with entangled heavy bird from dipping into mud or water (Figure 8.8).

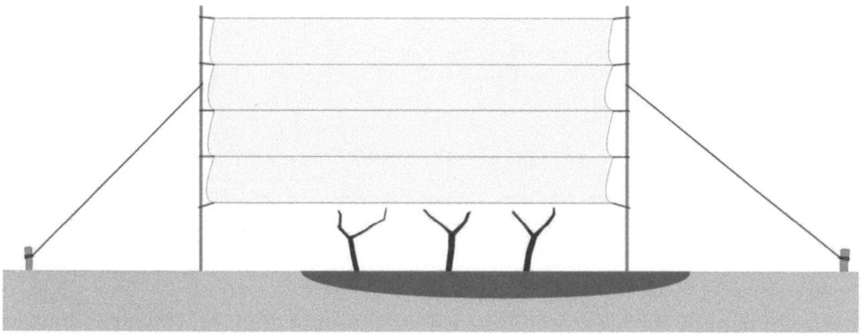

Figure 8.8: Mist nets with lowest shelf protected from dipping into water.

Wader mist-nets need very strong poles. They should be longer than those used for passerine catching, because waders are usually captured in muddy areas or over the water and sometimes more than 50 cm of the pole stay in the soft ground. To make the work easier, poles should be light-weight e.g. made of fiberglass. Telescoping construction makes these poles easy to transport and storage. Their colour should be dark (to blend into the habitat) at the netting site. The pole surface should be smooth enough to allow the net attachment loops to slide smoothly on and off the pole.

The selection of an appropriate mist-netting site is the most important determining factor for success. Knowledge about daily movements and activity patterns of target species is essential. The familiarity with preferred flight paths between feeding and/ or roosting areas is of utmost importance. Erecting mist nets at sites where the outline of the net is clearly revealed against a monotonous background, such as the sky or open water, decreases catching success. Clearings between vegetation or narrow spaces between two high banks (with a dark but variegated background) is an optimal netting site during the day. At night, mist-nets may be erected in flat open areas.

Multiple mist-nets are more effective than single ones. Formation of "L" or "V" shaped arrays increase capture rates. At night, especially when tape-luring is used, the most effective is array of a "T" shape, and when mist-nets are erected in large, flat feeding areas the array of letter "H" or double "T" (standing one on the top of another) is advisable. When waders are found alongside the coast, two nets (one above the waterline and one in the water perpendicular to this border) will be sufficient for catching (Figure 8.9).

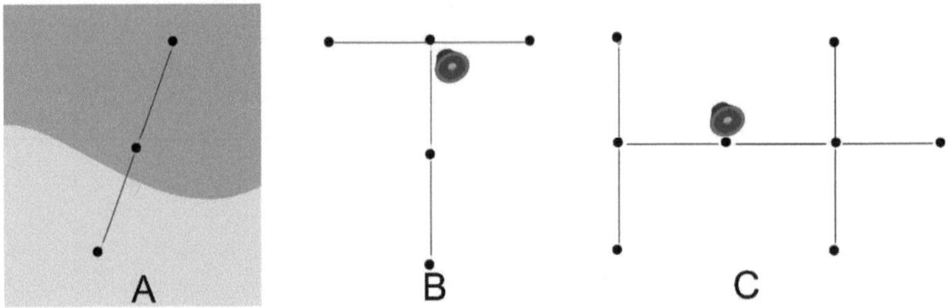

Figure 8.9: Different types of nets arrangements. A. At the border between land and water, B. and C. Overnight operation with tape-luring. Optimal location of loudspeakers is shown.

Tape lures usually increase results of wader catching at night. However, it seems that some calls work better than others. Thus, some experiments with voices obtained from different sources are needed. Clarity of the recordings seems to be very important when optimizing catching. This is not a matter of recorded voice, but also good quality of the audio system is important. Often, a breeding call is used to attract waders to mist-nest during spring and autumn migration (see Figuerola & Gustamante, 1995). It seems that breeding calls attract waders only just before and just after the breeding season and are ineffective the rest of the year. Each species reacts to its own calls, but, breeding voices of some species are quite universal. For example, breeding call of Black-tailed Godwit clearly increases catching results of ruffs and Eurasian curlews in Poland, while the voice of the Redshank attracts dunlins and knots in SE England (Clark & Austin, 2005; author's own data). The best results are obtained when a mixture of 3-4 different calls are recorded on a single CD track. In the case of species like Common Sandpiper, which gathers in flocks in the evening, a flocking call should be used instead of a breeding one. This kind of voice seems to be also the best option for catching waders in wintering sites, when birds do not react to breeding calls. Recording of large flocks of calling redshanks was also effective not only in England, but also in USA, where this species does not naturally occur (Clark & Austin, 2005).

In some cases, waders also react strongly to distress calls and quickly approach mist-nets during the day. It seems that the continuous distress calls caused such confusion that several birds would keep on flying around the recorder, increasing the chance of being caught. However, longer play back times did not lead to better results. When the tape is played frequently in the same area, most will get used to it, but newly arrived, inexperienced birds will react by approaching the nets (Haase, 2002). However, it should be noted, that use of distress calls not always leads to an increase in the number of caught waders, but results are less than what would be expected without a lure (Clark & Austin, 2005).

In the case of catching common snipes, all types of calls made by this species used simultaneously (i.e. drumming, displaying and alarm call) were quite effective. Attracted birds usually flew around mist-nets and then landed close to the speakers. Only few snipes were caught directly, most entered mist-nests when they were flushed in the nets by very quickly approaching people. The highest number of birds was caught in the morning at 5.00 - 9.00 a.m. (Pinchuk & Karlionova, 2006).

Different audio system may be used. Exploiting standard car CD-player with minimum output of 2 × 40W and two at least 80 W speakers is convenient, because such system is powerful and ready to use just after buying. The power of speakers should be at least twice of the player output to avoid sound distortion when system is working at full power. The player should be hidden in plastic box wile speakers in plastic bags to avoid moisture from the air. The standard small car battery (40 Ah) has enough power for working such system through whole night. Larger batteries work longer, but they are much heavier and limit the mobility of the system.

Cardboard silhouette decoys or plastic birds set underneath the nets appeared to attract waders. These silhouettes should be placed against water, which supported the best background at night. When catching far from the water at night, good results were obtained when an artificial pond was constructed near the erected mist-nets. It might be one plastic sheet (e.g. 10 × 10 m) lying on the ground with puddle in the middle.

8.2.1 Floating Mist-Nets

Other idea of mist-netting above water surface is to use floating mist-nets (see Pollock & Paxton, 2006 for details) (Figure 8.10). The most important parts of the floating mist-net setup are buoys, which are composed of floatation block, a length of aluminium conduit into which mist-net poles are dropped, and a weight suspended beneath to keep the buoy and pole upright. The heavier the weight used, and the farther the weight is attached from the floatation part, the more stable the buoys will be which result in much easier mist-net setup. Buoys dimensions present at the Figure 8.10 result in stable mist-net that could be used to safely capture the largest wader species with no risk of tipping. To keep the net under proper tension, it is necessary to tie each buoy and net pole to a stationary object. Moreover, to ensure that captured birds would be held well above the water, the net's bottom line should be placed at least 70 cm above the water.

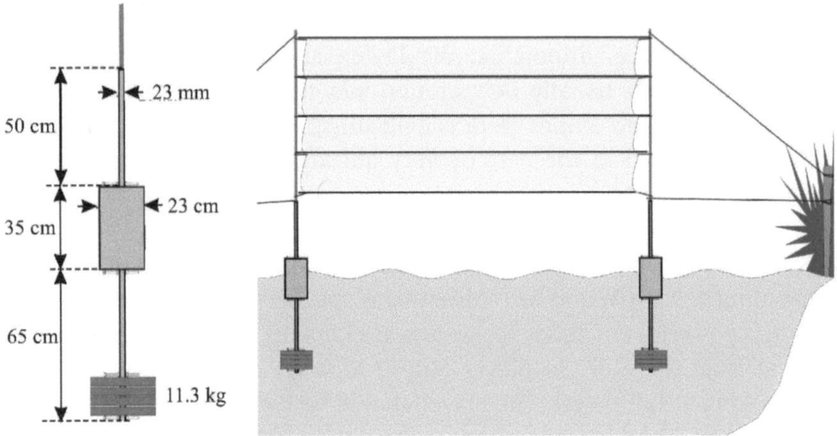

Figure 8.10: Schematic of floating mist-net with inset detailing the construction of a buoy (according to Pollock & Paxton, 2006).

8.2.2 Single Shelf Mist-Nets

In some areas, about 50% of mist-netted waders were caught in bottom shelf (Tree, 1972). In such cases, single shelf mist-nets may be used instead of larger ones. This kind of net is cheaper and much easier to operate (even by one person). In the case of low mist-nets, even slight rises in the ground provide good background for the nets, and that is why they can catch waders also during daytime. Single shelf mist-nets should be placed as low as possible, and in general over dry ground, not over the water. Many waders enter this kind of nets by walking and stay enmeshed in nets pocket. Similar to multi-shelf nets, low nets are effective when set at right angles to the shore and amongst higher vegetation patches in open wetlands.

8.2.3 Mist-Nets for Catching Waders in Intertidal Areas

In intertidal areas, mist-nets are erected at night in low tide. Increasing water levels force birds to escape from a flooded area and fly towards high-tide roosts. Thus, the best option is to set long lines of mist-nets perpendicular to the direction of birds' flight. Due to the rapid increase of water level, only one net checking is available and the team closing nets follows the people responsible for removing birds. Hence, birds must be taken from mist nets quickly and (to avoid cases of seriously entangled birds) the mist nets used for night catching in such areas differ from standard ones having smaller mesh (e.g. 19 mm) and thicker thread.

8.3 Leg-Hold Noose-Mats

Noose-mats are constructed with monofilament fishing line (nooses) attached to mats of wire mesh. Leading edges should be bent to eliminate sharp edges that might injure birds. The size of the mat might be different, however it is more convenient to use not too big ones. Mehl *et al.*, (2003) suggested sizes ranging from 30 × 75 cm to 10 × 90 cm. Fishing line should be thin (between 0.2 and 0.3 mm). The noose-mats are fairly cheap, but to construct a single mat, it is quite time consuming.

At one end of the each monofilament line, a single loop should be tied on, leaving about 3 mm small tab for quick removal of trapped bird. Free, unknotted end of the noose is then attached to the wire. The attached knot should be tightened by inserting a pencil through the noose and pulling away from the trap. The diameter of the fully open noose is about 5 cm. Nooses must stand upright, with the opening of the noose parallel to the length of the wire mat (Figure 8.11). Nooses should be spaced adequately to avoid gaps and overlapping between neighbouring nooses. In sandy substrate, the wire mesh should be submerged with only nooses exposed above the sand.

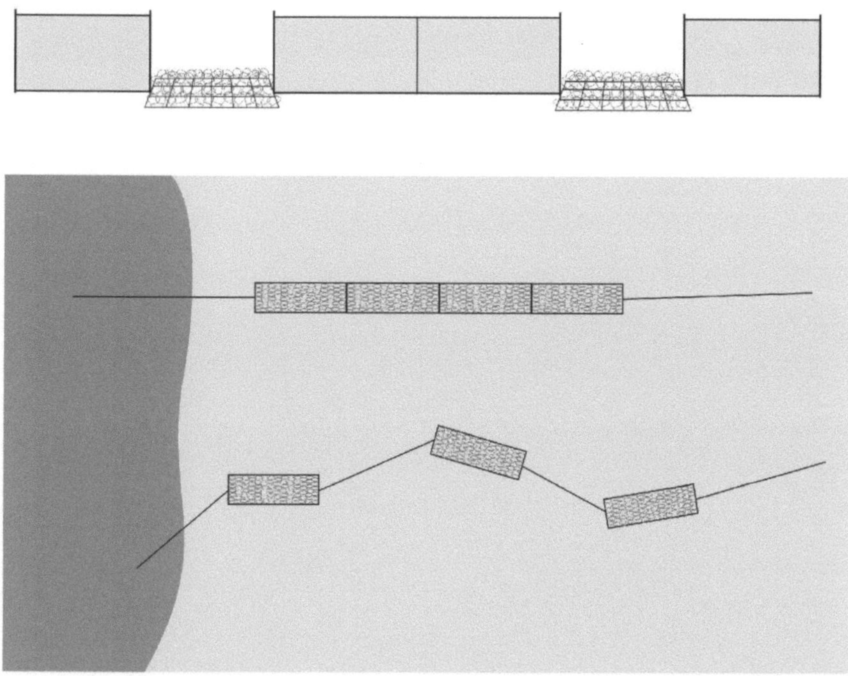

Figure 8.11: Set of guiding fences and noose-mats. Different placements of noose-mats and guiding fences (dark grey: water, light grey: sand or mud).

Guiding fences might be of the same construction (like in the case of walk-in traps with "legs" to put into the ground – see Figure 8.5). Waders generally foraged into the wind, and it is better to put sets of noose-mats with guiding fences perpendicularly to the wind creating a barrier across the site. Another possibility is to arrange noose-mats in V-shaped system, which is similar to the guiding fences of walk-in traps (Figure 8.11).

This kind of traps is lightweight and easy to transport and may be used on substrates with no dense vegetation at coastal and inland sites. It should be remembered that noose mats require careful handling and maintenance, because flattened nooses drastically reduce trapping success. Snared birds should be removed as soon as possible, so the one person has to stay all the time close to the noose-mats. Injuries of birds caught in properly managed noose-mats are rare (Mehl *et al.*, 2003).

9 Wader Station Laboratory Equipment

9.1 Wader Transport and Storage Devices

In the case of waders, bags might be used only to carry small species (e.g. Little Stint, Dunlin, Common Sandpiper), but not in the case of larger species (e.g. Bar-tailed Godwit, Whimbrel, Greenshank), which are sensitive to detainment. Even small waders cannot stay in bags and should be placed in storing containers after bringing them to the ringing site. It is much better to have containers, which may be used both for transport and keeping birds. For carrying most wader species, containers or wicker baskets of the dimensions: length 50-80 cm, width 30-40 cm, height 40 cm, should be used. For larger species the height of the container must be greater to allow them stay inside and instead lay on their belly (which prevents them from having leg cramps). For curlews, this height should be at least 60 cm, but in case of storing container even 1 m, in which bird can stand with headroom (Bainbridge, 1976; Stanyard, 1979). The walls of such containers must be airy, the floor hard and the “roof” covered with waterproof material; this prevents the birds from getting wet in rainy weather. The flooring should be cleaned from time to time, so it is handy if it can be removed. Such a container can be, for instance, made of a metal frame covered with fabric, with the removable floor made of plywood or plastic. Birds are removed through the entrance; this must be fastened and large enough to bring out even the larger wader species without any problems. If wicker, or plastic, boxes are used for carrying birds, one should remember that their tangle should be very thick (holes no larger than 1 mm). If holes are wider, waders may thrust their bills or toes into the holes, which may end up with a fracture or with the leg being sprained. Containers should be cleaned and washed regularly.

When the container is carried, it must not be shaken or waved. For that reason, holding the basket firmly in one hand is better than hanging it on the arm. Different species of similar size may be carried together, but the containers must not be overcrowded. The birds must have ample free space in the box.

For storage of waders (when there is a “ringing queue”), the same containers used for carrying are usually adequate. At the station, at least one larger container should be provided (where the birds may be put if numbers are very large) (Figure 9.1) and no less than 5-6 carrying baskets.

9.2 Ringing and Measuring Tools

Waders should be ringed with steel rings, and therefore, a set of special pliers for clenching rings should be available at the station. The most convenient arrangement is to have separate pliers for each, or two neighbouring ring-sizes with the holes for

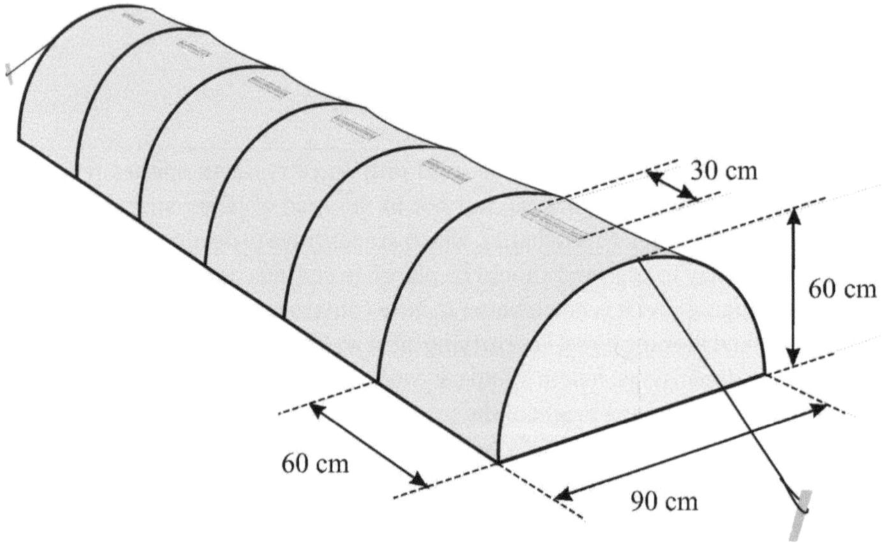

Figure 9.1: An example of keeping cage (according to Clark, 1986).

the ring placed close to the gripping arm. The shape of the hole should be slightly elliptical (Figure 9.2), this will slightly flatten the ring when it is clenched. It is important that the pliers open and close easily and have the right profile to clench the ring properly in only two moves: the first, closing a chink of the ring and the second, giving the right shape to the ring after turning it for 90° and finally closing the chink. The use of improperly profiled pliers may prolong the handling time two-fold!

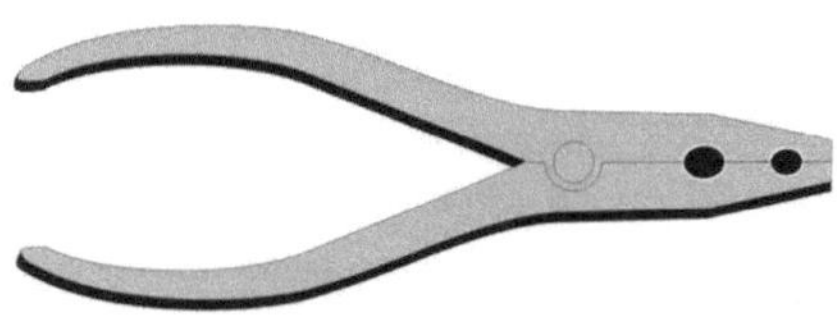

Figure 9.2: Pliers for closing steel rings on waders, note that there are only two ecliptic holes.

The Wader ringing station should be supplied with a set of two rulers, callipers and a digital balance for bird measuring. Due to the different sizes of waders, it is practical to have a small 30 cm ruler, which is enough for the majority of species up to the size of Oystercatcher, and larger ones, 50 cm for measuring the wing length of large species like Curlew and Whimbrel. A stop at the zero end of the ruler will facilitate the measuring of waders, since their wings usually are longer and more "stiff" than e.g. passerine wings. Dial callipers are much more convenient to use than vernier ones. The most often used callipers were made of plastic, which is lightweight and handy, but very sensitive to any (even small) particles of sand within racks. Thus, it is important to check them before measuring a proper set of zero points, and clean the racks of the callipers when necessary. After few seasons, moving calliper arms becomes "softer", which may lead to errors in reading the result of measurement. Therefore, it is better to replace callipers every 2-4 years of use. With standard callipers, the lengths up to 150 mm can be measured. This kind of calliper could be modified by adding a special block to the fixed arm (Figure 9.3) (Green, 1980). It makes positioning of the occipital part of the head in calliper arm much easier. For larger, long-billed waders, standard 150 mm callipers is not enough; however, it is not easy to get larger ones. Reynolds (1986) showed a device which may be useful in measuring the total head length of long-billed species (Figure 9.3). This is a stepped ruler with a basal section cut away for bird head. This device allows measurement to the nearest 1mm instead of 0.1 mm offered by callipers. However, in the case of Eurasian Curlew (with total head length ranged between 130 and 190 mm) the accuracy of 1mm is quite enough. An electronic balance with accuracy of 1 g and maximum capacity of 1 kg is enough for all wader species. It should be noted that some cheap models may be sensitive to low voltage of batteries and could show weight erroneously.

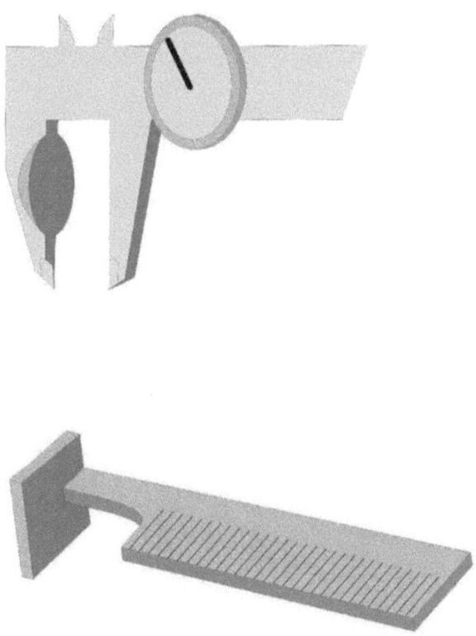

Figure 9.3: Modified callipers for measuring total head length and stopped ruler for measuring total head length of long-billed wader species.

10 Wader Station Laboratory Methods

10.1 Standard Set of Measurements and Scores

The **total head length** (Figures 10.1-10.2) is measured with callipers. The easiest way is to use the wider part of the jaw for this task. The bird's bill is held close to the tip. The calliper's inner jaw is placed at the tip of the bill and kept in place with the thumb and the second finger. The outer calliper jaw is then pressed against the most exposed part of the occipital bone. Slight movements of this calliper will help in finding the right position. The axis of the bird's head should be parallel to the calliper, and it is important this angle is maintained while taking the measurement. Use only sufficient pressure on the calliper to press the contour feathers to the skull, and never curve the tip of the bill. In the case of short-billed species, be careful not to block the nostrils! When learning to take this measurement, it is recommended first to find out the most exposed part of the occiput with one calliper jaw, then open the other jaw carefully (the one fixed to the occiput stays in place) and find the correct length by moving it back to meet the point. With this procedure, the proper pressure of the callipers is easily attained. Modification of fixed arm of the callipers (Figure 9.3) makes this measurement much easier. For species with total head length larger than the standard callipers range (150 mm), special measuring device may be constructed.

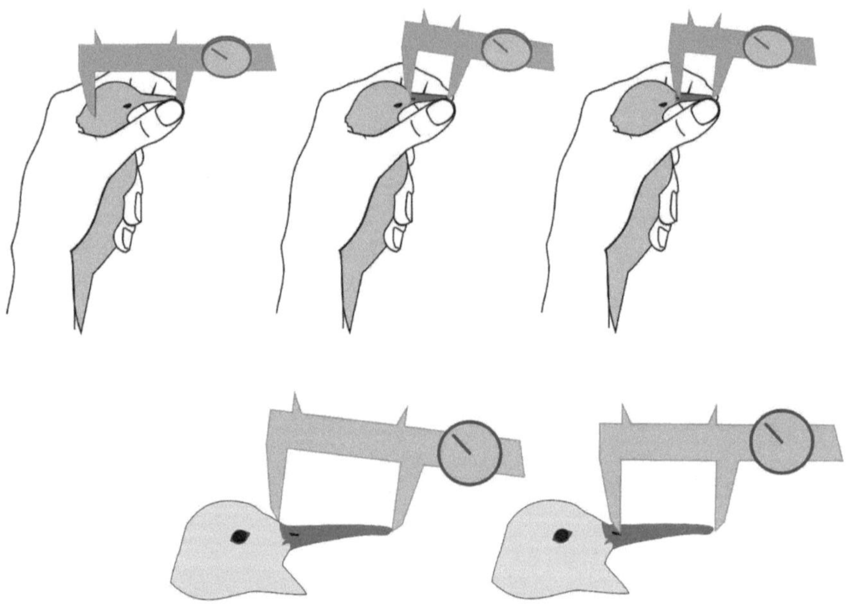

Figure 10.1: Technique for measuring total head length, bill length and nalospi with callipers (consult standard descriptions in the text).

Figure 10.2-1: Bill-length measurement in waders. Photo W. Meissner.

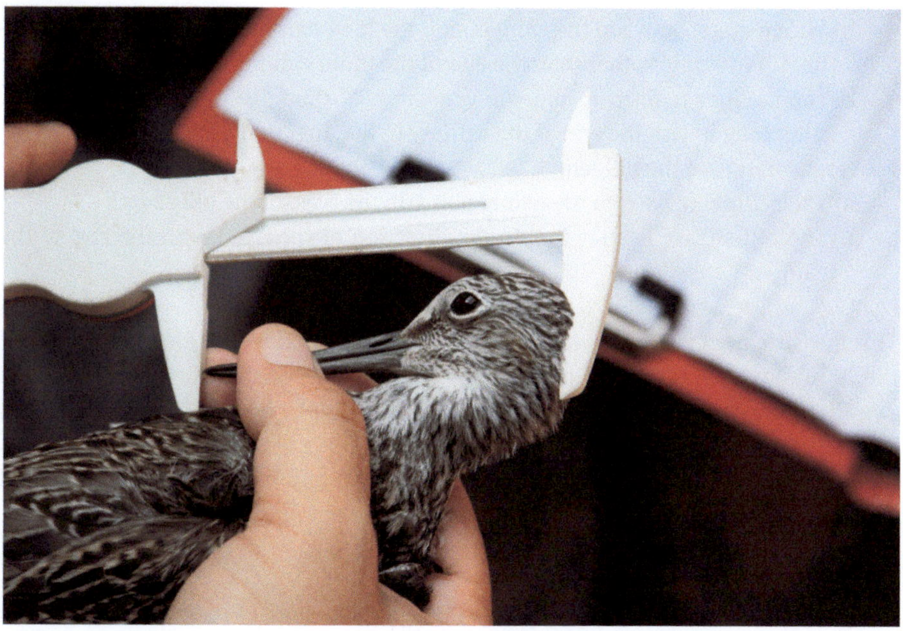

Figure 10.2-2: Total head-length in waders. Photo W. Meissner.

The most frequent errors include:
1. Underestimation of the head length
   a) Extremely strong pressure of the callipers causing change to the shape of the bill tip
   b) Forming the wrong angle between the calliper and the long axis of the bird's head
   c) Contacting wrong point between the calliper and the occiput
2. Overestimating the head length
   a) The calliper is not close enough to the occiput
   b) The calliper is fixed to the muscles of the upper part of the nape

The **bill-length** is also measured with callipers (Figure 10.1-10.2). The bill should be held the same way as when measuring the head length, but the tip of the bill is placed at the tip of the calliper's jaw, not the wide part. Then find the most distant point of the edge of the bill with the tip of the outer jaw (Figure 10.1). Keep in mind that part of the bill's sheath is very elastic in some species, so be careful to take the measurement in its natural position, and avoid over-tightening the calliper.

The most frequent errors for measuring bill-length include:
1. Underestimation of the bill length
   a) Extreme tightening of the calliper to the bill, causing shape of the bill tip to change
   b) Wrong angle between the calliper and the long axis of the bird's bill, fixes the bill tip to a point other than the end of the inner calliper
   c) The calliper is fixed to the wrong section of the edge of the bill sheath, or stretching it with the end of the calliper in the direction of the bill tip
2. Overestimation of the bill length
   a) The calliper is fixed to the wrong point on the edge of the bill sheath
   b) The bill sheath is stretched in the direction of the head with the end of the calliper

This measurement is difficult to take accurately in species where the border between the horny and the feathered parts of the bill is poorly demarcated, such as most species in the genus *Tringa*. In these species, the distance from bill tip to the nostrils can be measured (nalospi) instead of bill length or the measurement can be left out. There is a strong correlation between the bill length and the total head length, and in most cases, an analysis of both these parameters in individuals whose bills have finished growing gives similar results.

Wing length – A stop at the zero end of the ruler will help when measuring waders' wings, which are usually longer and stiffer than passerines' wings. The folded wing should be held parallel to the body on the ruler. The carpal joint is placed on the ruler's stop. Use the thumb of the same hand to press the wing firmly against the ruler. Use the thumb of the left hand to straighten the primaries to their maximum length by pressing down the curved wing while applying slight lateral pressure towards the bird's body at the level of the primary coverts. The third and fourth fingers of the left hand control the folding and straightening of the wing (Figure 10.3).

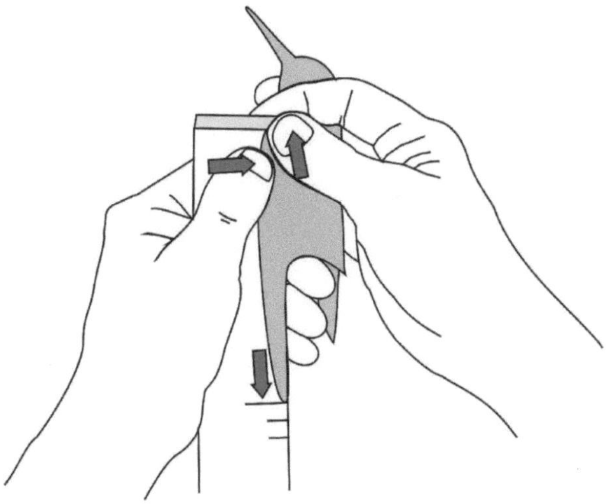

Figure 10.3: Technique for measuring wing-length in waders.

The most frequent errors:
1. Underestimating the wing's length by:
   a) Not pressing the wing fully against the ruler
   b) Not fully straightening the primaries
2. Overestimating the wing's length by:
   a) Not holding the carpal joint firmly against the ruler's stop
   b) Not fully bending the carpal joint

Length of tarsus + toe (without claw) is measured using a ruler with a stop. The tibiotarsus (tibia) is pressed against the stop and the tarsometatarsus (tarsus) to the surface of the ruler. Push the tibia to the stop with the fingers of the right hand and straighten the tarsus and the longest toe along the ruler's surface with the fingers of the left hand (Figure 10.4-10.5). Both parts of the leg should be pressed against the ruler so that the correct angle between them is maintained. Ensure that the whole toe is held against the ruler. Do not include the claw when reading the value.

Figure 10.4: Technique for measuring the length of the tarsus + toe.

The most frequent errors include:
1. Underestimating the length of the tarsus + toe by:
   a) Maintaining too small of an angle (acute angle) between the tibia and the tarsus
   b) Pressing the toe firmly against the ruler
2. Overestimating the tarsus + toe length by:
   a) Maintaining too large an angle (obtuse angle) between the tibia and the tarsus

Data on the moult index of primaries should be collected in standard way described in *Special Studies* section.

10.2 Additional Measurements and Scores

Tarsus-length is measured with callipers from the notch on the back of the intertarsal joint to the end of the tarsus bone (Figure 10.5). The toes should be bent about 90° to the tarsus. It is recommended to use wider part of the jaw for this task.

Nalospi, the distance between the tip of the bill and the proximal edge of the nostrils, is measured with callipers (Figure 10.1). This measurement is recommended for species in which the border between the horny bill sheath and the feathers on the head is poorly demarcated. It is strongly correlated with the head length. Some difficulties might occur while holding the calliper against the proximal edge of the nostril. The bill's sheath is more supple at this point and is easily deformed. When the bird has narrow nostrils it might be difficult to find the right place to hold the calliper.

Figure 10.5-1: Measuring the tarsus-length in wader.

Figure 10.5-2: Measuring tarsus + toe.

Third primary length – see *Additional Measurements and Scores* – p. 97.

Fat score scale for waders. This fatness scale described by Meissner (2009) is based on determination of the amount of fat in the axillary region and only then, if necessary, in the furcular region (Table 10.1, Figures 10.6-10.8).

Table 10.1: Description of fat classes in waders.

Fat score	Axillary fat depot	Furcular fat depot
0	All area under the skin is flesh-coloured with no light fat traces. Sometimes reddish traces could be visible, but not yellowish.	Not necessary to check.
1	Small, yellowish fat patch. Usually elongated, filled depression between two muscles.	Not necessary to check.
2	Fat patch overflows depression between muscles, is more rounded, but still flat, (not convex!); does not fill the depression, but usually reaches the depression edge.	No fat visible or small traces without clear layer.
3	Fat patch overflows depression between muscles, is more rounded, but still flat, (not convex); does not fill the depression, but usually reaches the depression edge.	Fat layer fills the furculum usually up to 1/3 (clearly less than 1/2, and might rarely be more).
4	Fat patch forms a convex cushion and at least partly fills the depression. Muscles are still visible around the fat patch.	Fat fills about 1/2 furculum; edges of fat patch may reach clavicles.
5	Fat patch forms a convex cushion and at least partly fills the depression. Muscles are still visible around the fat patch.	Fat fills the whole furcular depression and its surface is not very concave. In the middle of the fat pad, a small depression may be seen.
6	Fat patch forms a convex cushion and totally fills the depression. Muscles are not visible within apterium.	Not necessary to check, but at this stage fat pad in furculum is convex or at least flat with no depression.
7	Fat patch forms a convex cushion and totally fills the depression. Muscles are not visible within apterium.	Fat patch fat overflows furcular depression and reaches abdominal region.

Good lighting conditions, especially when capturing birds at night, are important for proper assessment of the fat score. However, the most common mistakes result from improper handling of birds. Incorrect neck placement when assessing fat deposits in the furcular region can influence results, especially when distinguishing between fat scores 4 and 5. Excessive neck extension causes underestimation of fat present, whereas neck constriction causes overestimation. Assessing amount of fat in axillary region seems error-proof because differences in how birds are held have no apparent effect on estimates, if the whole apterium is visible and the wing is pulled away.

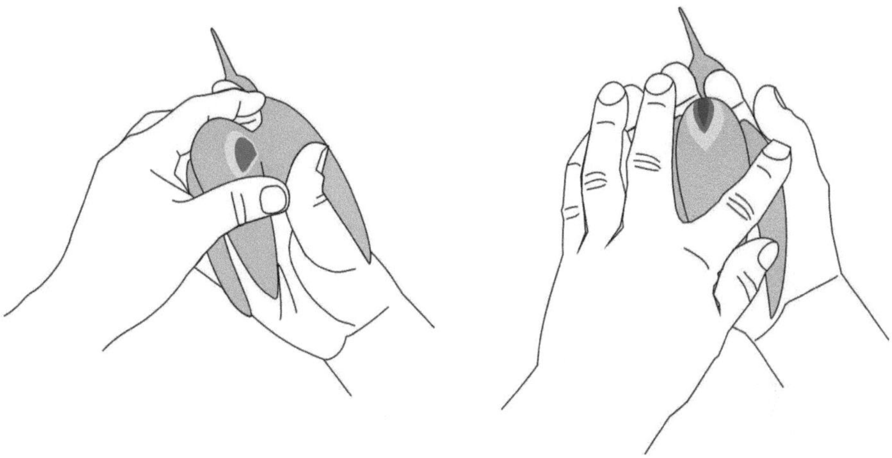

Figure 10.6: Technique for holding a small shorebird while assessing fat in the axillary region (left) and in the furculum (right).

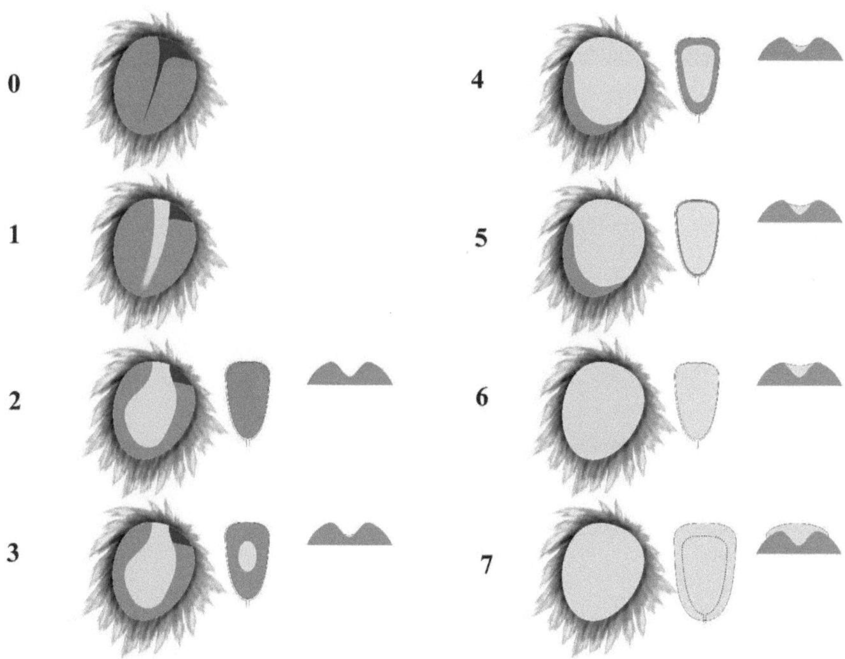

Figure 10.7: Visual representation of the amount of fat accumulated within the axillary region (images in left column) and in furculum (ventral view and cross-section; images in right column), with associated fat scores. White represents fat, light grey indicates muscles, and dark grey represents the depression under the arm ("hole") and along the breast muscle. Detailed description in Table 6.2.

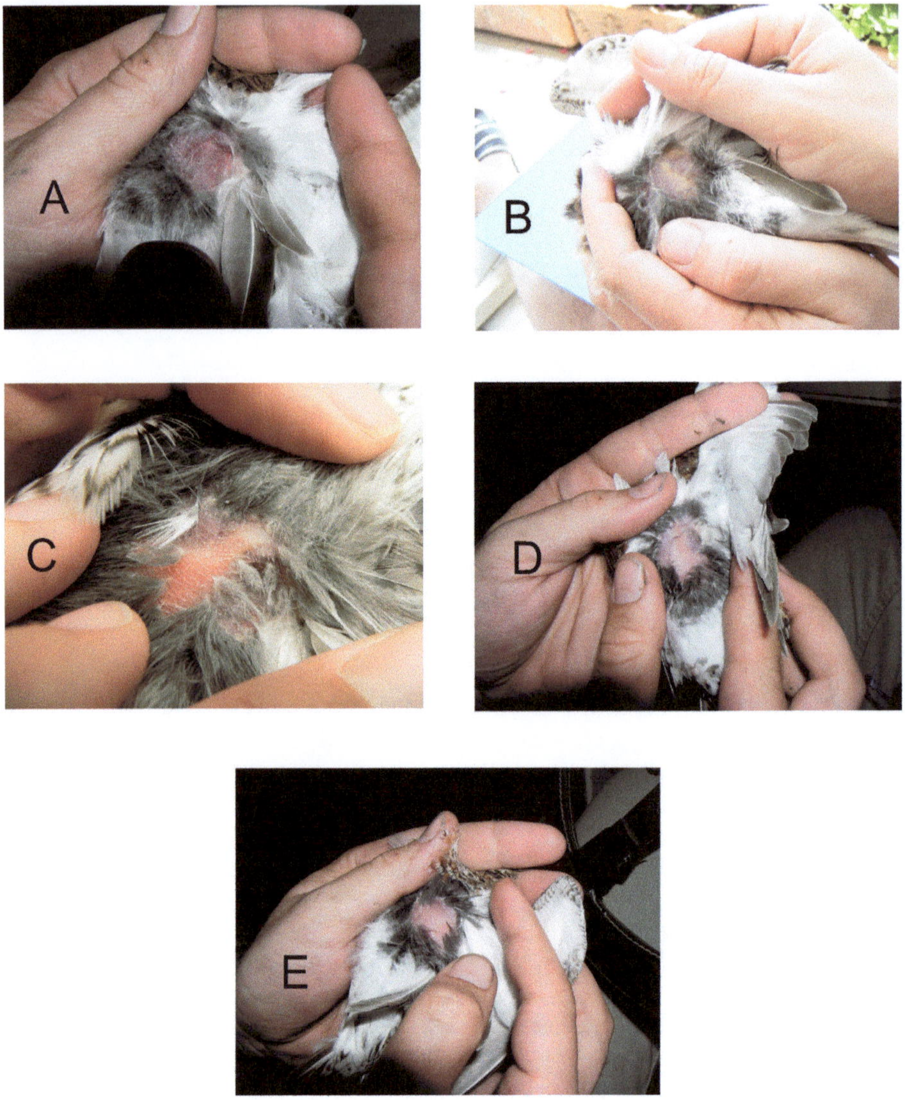

Figure 10.8: Examples of the fat scores: A. Score 0, B. Score 1, C. and D. Scores 2-3, E. Scores 4-5. Photo W. Meissner.

Distinguishing between fat scores 1 and 2 or 2 and 3 can sometimes be difficult and different observers may assign different scores for a particular bird. However, such discrepancies do not exceed one degree of fatness. For larger species, like curlews and oystercatchers, scoring fat in the furculum is difficult because of their very deep, interclavicular depression. As a result, distinguishing between scores of 2 and 3 may be impossible because amount of fat in the furcular depot could be difficult to see and, for these species, these two fat classifications should be combined.

Muscle score. The main idea of this score was described in the *Additional Measurements and Scores* section, p. 97. However, in the case of waders, especially, species larger in size, breast muscle contour could be measured by pressing and moulding a piece of plastic-coated copper wire with an outer diameter of 1-3 mm against the contour of the breast at the midpoint along the length of the keel. This technique was developed for poultry by Gregory & Robins (1998). The arc formed by the piece of wire may be traced onto paper and it should help in distinguishing between subsequent muscle scores.

Amount of protein stored in pectoral muscles may be assessed more precisely using quick-setting alginate gel (Selman & Houston, 1996). Briefly, the alginate is mixed with water to give a smooth, runny paste, with which a plastic tray was filled to a depth of 3 cm. The gel remains fluid for 2 min before hardening within the next 30 seconds. A bird is placed breast downwards in the gel, 2 min after the gel was mixed. When the gel had solidified, the bird was lifted clear of the mould. Alginate does not adhere to the feathers, and the process causes minimal stress to the bird because it was held for less than 30 seconds. Once the bird was removed, the alginate retains an exact replica of the shape of the pectoral muscle region. This is then cast permanently in plaster of Paris. The plaster casts of pectoral muscle profile are then cut dorsoventrally by a band saw at the midpoint between the furcula and the posterior end of the sternum. The cut surface is then placed downwards on cardboard, and the breast outline is traced giving the area to be measured. This area shows strong correlation with pectoral muscle lean dry mass and may be used as an index of protein reserves.

Wear categories of primaries and secondaries. The timing and pattern for the moult of flight feathers helps in the ageing of waders and also in recognising different populations, which have different moult schedules. The categories of wear described by Prater *et al.* (1977) are widely used (Figure 10.9). A magnifying glass is helpful when distinguishing among these categories. During the field study, the numbers from 0 (fresh feather) to 3 (very worn feather) could be attributed to subsequent wear categories and it is convenient to use notation as in the case of moult index of primaries.

Example: $\mathbf{3^2\,2^6\,1^2}$ equals to:

- Two outermost primaries: very worn (3)
- Next six primaries: worn (2)
- Next two primaries: slightly worn (1)

Note: that the sum of exponents must be equal to the number of primaries (10 in case of waders).

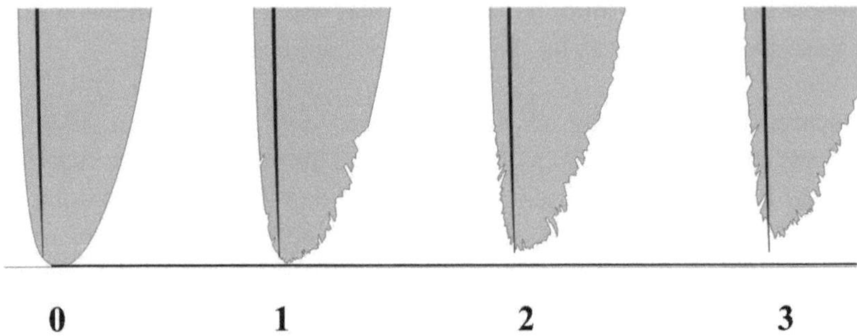

Figure 10.9: Wear categories of flight feathers.

Features relating to the plumage colours. Colours of the plumage can help to distinguish different populations of waders. For example, in the Bar-tailed Godwit and the Whimbrel, a relationship has been established between the colour pattern of the under-wing scapulars and the breeding area. A similar key has been devised for Dunlin that notes the extent of white reaching the shaft on inner primaries and the presence of median wing coverts of the adult-buff type. Devising a scale to note these types of features starts with a description of the range of variation in a feature and goes on to define a succession of criteria, allowing a distinction to be determined between each category. Digital photography may be used for registering such scores. However, images recorded by digital camera are not only dependent upon the characteristics of the photographed bird and the ambient light, but also upon the characteristics of the equipment. Thus, caution should be implemented and suitable calibrations developed before such investigations are undertaken. This issue was discussed in detail by Stevens *et al.* (2007).

11 Wader Station Laboratory Working Routine

The routines and rules of the wader ringing station are very similar to those of the passerine station. Attention should be paid to recommendations resulting from different trapping methods and a slightly different set of measurements.

When at the station, species vulnerable to a long retention should be ringed first, i.e. godwits, curlews, whimbrels, greenshanks and spotted redshanks. If any passerine were brought, e.g. wagtails, they should be ringed in a second turn. After ringing, the age and, and if possible, the sex of the bird should be noted together with the ring number. This usually requires opening the wing, and so the moult score and other notes referring to the wing may be taken at the same time, e.g. features connected with the colouring of feathers and the wing length can be measured. Once the ruler is held in hand, the next measurement taken should be the tarsus + toe length. After that the callipers are produced in order to measure total head length, bill length and possibly other measurements taken with this tool (nalospi, tarsus). Finally, the bird is weighed.

In order to shorten the time of retention, the best thing is to work with a 3-4-person team while ringing. One person takes birds out of the basket and clenches the ring, another takes all the measurements, a third weighs the bird and releases it, while the fourth one takes notes; alternatively, the third person takes notes while not weighing birds). When releasing waders in the daytime, they will fly away by themselves. Larger species may be carefully thrown in the air, taking care that there are no bushes or lines in their way. At night a bird taken out of the laboratory should be given time to adapt to darkness before being released. This lasts longer than in the daytime, and therefore the person weighing birds cannot release them at the same time. The bird must, under no circumstances, be allowed to leave the ringing station "on foot". If it is unable to fly but can walk and is not injured, it is better to bring it to the feeding place and release it there. After some time, most of these "immobilized" birds will fly again.

12 Wader Counts

The phenology of wader migration at stopover sites is the subject of many studies, and forms the basis of other more detailed analysis. In passerine ringing stations, data from catches are frequently used to illustrate migration dynamics or monitor population numbers. In the case of waders, the size of discrepancies in migration patterns obtained according to counts and trapping in walk-in traps differed between years, and it was related to neither the aggregate number of migrants counted nor the total caught. However, it would seem likely that in the case of a very rare species, the dates on which birds are trapped would show little correlation with migration pattern (Meissner, 2008). It seems that the main drawback of using data from birds caught in walk-in traps is that trapping efficiency may differ not only between years, but also within a season due to many external factors (e.g. weather conditions, water level changes). Therefore, catching data should only be used with caution in studies of migration phenology in waders. On the other hand, it should be kept in mind that during the count, we do not know the extent to which we deal with individuals counted on a previous day when birds are counted every day.

The proportion of waders that are juveniles is often considered a good indicator of breeding success. Two methods can be used to obtain such data: counting or catching. However, catching methods may be particularly prone to bias, mainly toward juveniles due to their lower awareness and greater naivety. Moreover, depending on local conditions, the age structure of waders caught at least in walk-in traps may be biased towards juveniles or adults (Meissner, 2007). Therefore determining the proportion of juveniles from regular counts may be the better method of assessing wader productivity at stop-over sites during autumn migration than catching. However, the age structure monitored in any given season also partly depends on the length of birds' stopover at the study area, and the duration of stay may, and will, differ between adult and juvenile birds.

Wader counts should be conducted regularly in a standardized way and cover a large enough area so that bias arising from systematic differences in the distribution of adults and juveniles is minimised (Harrington, 2004). However, catching remains the main method for recording productivity in most species in winter (when the differences between adult and juvenile plumage become difficult to discern in the field) and in autumn (for species with only slight age-related differences in plumage).

To collect data on migration phenology or the changes of migrant numbers, every-day counts of waders resting and foraging in the surroundings of the station are recommended. If possible, at least two counts within every five-day period should track changes in bird numbers during migration. When waders are counted, adult and juvenile birds should be noted separately, if possible, in a given species. However, in stop-over sites with high turn-over rate, the number of caught birds may exceed the maximum number of seen during the day. In such cases, results of counts and catching should be analysed together. It is good practice to note the number of birds also during trap checking and other observations made during the day. It gives data on species, which appeared solitary or in very low number that might be not seen during the standard count. Only people that are capable of recognizing all species in the field and estimating the size of flocks should record counts. The best time for counting is around noon, after a few birds walk into the traps, and the majority of them are resting, which also makes counting easier. Remember to add the number of birds caught in the traps to the overall count! Many wader species have a tendency to prefer one particular type of habitat. For instance, e.g. curlew sandpipers or bar-tailed godwits are seldom met in a wet meadow, while, for example Sanderling and Turnstone practically always feed on sandy beaches. Therefore, it is preferred if the area where birds are counted includes information about all the habitats found in the immediate vicinity. A strict definition of the limits of the census area is important as well. The area should, if possible, have natural demarcations, and a meadow or a muddy bay should not be divided. On the other hand, the area must not be too large, because walking around it and counting the birds should not take more than 1-2 hours. If the count is done by a ringer, it should start immediately after all birds from the previous control walk have been ringed, so that birds brought from the next control walk will not have to wait for the ringer to return. A light telescope is very helpful in counts. It should be remembered, however, that some species (snipes) are very shy and spotting all individuals in the vegetation is practically impossible with a telescope. Therefore, places where snipes congregate should be walked over in a zigzag pattern, trying to flush all individuals on the wing.

Recently many colour-ringing schemes have been implemented for waders. It is worthwhile to pay attention to birds with such rings during the counts.

PART III: **General Issues**

13 Training Beginners: Bird Measurements

Measurements are of value to science when they are reproducible, i.e. do not depend on the individual characteristics of the person. The compatibility of measurements does not imply that two people present identical results for each bird measured independently by them, but rather, it implies statistical concordance for a series of measurements done by a number of people measuring the same sample of birds. This situation is attainable when the standard techniques are carried out strictly according to the rules described. The system of instruction must guarantee correct interpretation of these standard descriptions of techniques, ensure correct execution of measurements and cross check the results of this instruction.

Measurers are trained step by step as follows:

1. The measurer reads the descriptions of the standard technique, and the instructor explains them in order to cover any doubts that might ensue.
2. The instructor demonstrates the correct way of making measurements (slowly, with comments) and provides specimens showing typical fat deposit patterns. This step is complete on a small number of birds.
3. The measurer practises their fresh knowledge on a few specimens, under the guidance of the instructor. The goal is to attain measurements matching those of the instructor, who has measured the bird beforehand. Measure about 20 birds in this phase.
4. The measurer independently measures birds previously measured by the instructor (about 100 birds), then compares these results to those already noted and corrects their errors. When more regular errors are perceived, the measurer should discuss shortcomings with the instructor.
5. Initial checking:
   a. The beginner measures a series of about 50 specimens independently and without recourse to comparative measurements. His results are noted separately on a training chart. This series should involve birds of comparable size. The optimal, full control involves birds of the Goldcrest, Great Tit and Thrush size in order to detect possible size-dependent errors in measurements;
   b. After completion of the series, the beginner's results are compared with the correct measurements and all deviations are noted using coloured numbers at the corner of every line on the chart where results differ. These deviations are summarised algebraically for every parameter separately and the mean deviations are calculated. Deviations of less than 0.2 are treated as a correct result;
   c. If a greater mean deviation occurs, the instructor should carefully check the probable reasons for incorrect measurements, repeat stages 3 and 4 of the learning process and superintend until reproducible results have been achieved.

6. Final check:

Following a positive initial check, the beginner should measure some hundred birds alone and the instructor's checking, as under point 5, is repeated. It is worthwhile to check the beginner in the following season when his performance should have stabilized, or individual divergences will have emerged. The check is best performed if all the beginners of the research group can be compared with the "group-standard" or even the "country-standard" person.

Later, during work, there is a good custom to make comparisons between ringers when changes of measuring person occur: both persons should double measure some individuals or measure some, but more, individuals from the same group of migrants. The other solution, if there is no possibility to stay at the ringing station for a few days, is to compare measurements *ex post* on graphs made from the data collected in subsequent days as it was shown at the original graphs. In this case, use data about the most common bird is recommended.

14 Bird Mortality and Welfare

Migrating birds are evolutionary adapted to overcome migration barriers, compensate for losses of fat deposits, and endure stress from being exposed to raptors in alien and unfriendly habitats, etc. In the cases of birds being caught, the ringing station and we, the ringers, constitute an additional stress factor, and contact with ringing activities adds to the risk of dying before the next breeding season, when the bird has a chance to pass its genes to the next generation. At any rate, a majority of all birds present in late summer, in passerines as much as to 85%, will die on migration and in the winter season (Newton, 2008), irrespective of their meeting ringers or not; they would be condemned to death within some hours, days or at least months (Figure 14.1). In contrary, usually 99% of caught birds fly farther without too much disturbance (Figure 14.2). These statements give some distance to very emotional and sometimes hot discussions on the problem: are the catching losses in birds acceptable? Let's face the truth, and try to discuss this very serious problem in more detail.

Mortality among birds caught by bird ringers has many different objective causes.

Some birds die or are injured due to faulty catching technique, some are killed by raptors when caught and unable to escape, some die during the ringing process or soon after being released. This mortality, however, is not reason enough to stop ringing or other studies where caught birds are involved. We must keep in mind that

Figure 14.1: Natural death during migration. Most of migrants (up to 70-85%) die during migration as a result of natural causes (here lack of energy stores during flight over a desert). Aswan, Egypt. Photo P. Busse.

Figure 14.2: Most of migrants caught and ringed are still full of vigour and willing to fly farther. Photo P. Busse.

ringing results may and will save many more birds in giving some advice on how to protect birds and their environments more effectively. At the same time, the fact that there is unavoidable mortality connected with ringing presents a strong obligation on ringers to make every effort in order to reduce losses among birds in their custody. This is the main reason why bird safety is a main topic in this Manual. The preventive theme is given particular emphasis in the *Alarm Routine* section. In the following section, a few more general comments are given, and a short summary collects advice presented in various parts of the text.

14.1 Catching Devices

The methods used when catching birds for scientific purposes are generally safe for the birds. However, "generally safe" does not mean that there are no birds injured or dead because of the catching device.

(a) **Nets:** Nets made of thin thread are intended for smaller species; when they catch larger species they are more likely to cause injury than nets made of thicker thread. The most common injuries are scars to the skin and cut-off tongues. Cases where birds hang themselves are comparatively rare; these occur in strong wind when birds are lifted out of the shelf and one mesh is pulled tight around the neck. **Remedy:** use thicker thread for standard catching.

Note that it is much safer for waders to use thicker thread for standard catching. It seems as if waders are more susceptible to injury in mist-nets than passerines living in woods or bushes. The latter are used to being rubbed or bumped against leaves and branches, and their bodies are stockier. Waders, on the other hand, are adapted to open and flat spaces and to running, but not to perching on branches. That is why they are more "soft-bodied", and therefore also more susceptible to injuries from the impact with mist nets (Meissner, 1992). Moreover, they fly with greater speed than passerines and the thin tread of wader-nets is more likely to cut the skin than the net designed for passerines. Still, mist-nets are very useful for catching waders, and in some areas, they are the only method available. Even at the ringing station, where the walk-in traps are the main catching tool, mist-nets may be used as an additional tool e.g. for catching birds attracted by tape luring at the night-time. When mist-nets are used to catch waders, they should preferably be watched continuously, and birds caught should be collected immediately. Old-fashioned, light-collecting binoculars will allow the ringer to check nets from a distance! Unexplainable cases of death are exceptional, although they occur where the bird hits the net and falls down into the shelf as is no longer alive. Sometimes, however, it may still be alive, but one could say it has fainted. This is probably caused by a psychological shock similar to that occurring in human life. The fainted bird may suddenly fly away – so do not put a bird apparently dead in your pocket! **No remedy:** accidents of this kind are unavoidable.

(b) **Walk-in traps:** Birds already caught in traps are vulnerable to raptors, since the latter learn very quickly how to exploit this source of prey. Mammals will prey on trapped birds mainly at night or at dawn, whereas birds of prey or Corvids and Gulls will be active in the daytime. When raptors begin to penetrate the traps, night catching must be abandoned. In the daytime, the traps may be guarded, and raptors are shyer than waders. In addition, a special spring-trap on the roof of the walk-in trap will act as an effective deterrent. A bird of prey caught in that way should be removed immediately and transported at least 10 km from the ringing station. A good effect is obtained by placing additional fences along the walls of the capture chamber. They should be mounted with a slope and fixed to the trap. A second solution is to drive long twigs forked at top into the ground; this will make attacks by bird of prey from the air difficult (Figure 14.3).

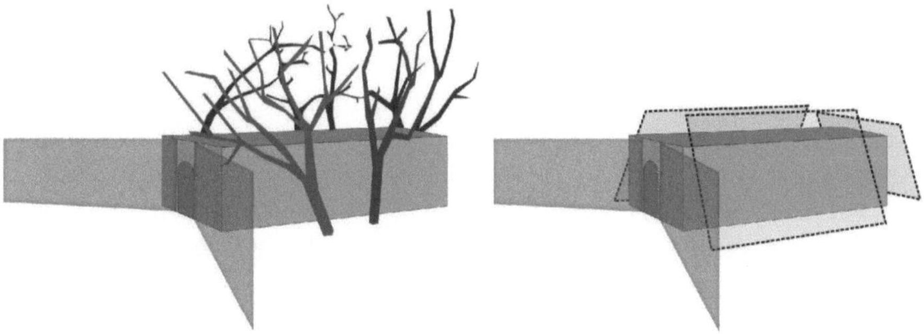

Figure 14.3: Protection of wader traps against winged raptors.

An electric fence positioned around catching site is very efficient against mammalian predators. When walk-in traps are placed along the coast, the total length of such fence becomes very high. However, the way of electric fence placement depends on the waterline shape and in some cases it is easier to separate the whole peninsula by short fence, than put long fence along the coast. The height of the fence should not be less than 50 cm with 8-9 wires between poles. The lowest wire might be unplugged, to allow passage of small animals. However, even a 50 cm high fence might be jumped over by foxes or feral cats. In such cases, a higher fence or two parallel lines of 50 cm high fence spaced apart about 1 m should be used.

(c) **Heligoland traps:** In the big Heligoland traps made from soft netting (non-metal net), a bird may get stuck and strangle itself in the top section of the trap; the dimensions of the construction and the fragile walls in many cases render all rescue attempts futile. Since accidents of this kind are more probable when the netting is broken, loose, or incorrectly fixed to the construction lines, regular maintenance of the trap and stretching of the netting will serve as an at least partial remedy against these losses.

In Heligoland traps with a terminal box, where birds fly against a glass pane, death as well as injury of birds entering the box at high speed occurs relatively often; blood effuses into the brain or eyes. Such constructions should not be used, or minimally, and the possibility of hitting the glass at high speeds should be reduced.

During mass trapping, the number of birds simultaneously collected in the final box should be limited by more frequent controls.

14.2 The Catching Process

All birds caught are readily accessible to raptors, and this is when the bird is hanging in a net or caught in a walk-in trap.

Avian raptors (especially the Sparrowhawk and Great Grey Shrike) hit the birds in the nets frequently. Local individuals are especially dangerous, as they learn quickly that the birds in the nets could be a source of food. However, they also learn quickly, that they can be caught, too. **Remedy:** there are limited chances to reduce such losses. In extreme cases, transport of a local individual far off the station area could be undertaken, if the bird will be caught. From the other point of view, however, these raptors naturally kill some birds to survive, so by netting potential victims, we make their hunting easier. In any case, some bird individuals are killed. During spring, migration of tits, especially great tits, in poor condition may kill other birds caught in the nets, sometimes, birds the size of thrushes. The brain of a killed bird is eaten first, followed by the rest of the body, and the ringer will find only bones and skin in the net. Tits are often caught when killing other birds and they usually continue to eat after being caught. **Remedy:** there is no possibility to eliminate such cases; the only action, which might reduce the losses, is not to remove killed birds as long as there is something to eat on them; new tits will not kill new individuals but clean the old corpses. This is a difficult strategy when visitors come to the catching area, however. Sometimes feeding the hungry birds with tallow from your butcher will help, but usually feeding is not overly effective since the birds are not residents of the catching site.

The most common wild mammal raptors killing birds in nets and traps are foxes, raccoons and different marten-like species. These animals are most active during night time, therefore they are most likely to attack owls and night migrants landing in the catching area early in the morning. Sometimes, however, specialized individuals of Marten and Ermine will attack and kill birds in broad daylight; Marten may develop into a sourge in reedbeds. **Remedy:** in practice there is no effective remedy against these raptors where mist-nets are used; in some cases, blood flour, used to intimidate stray dogs or spraying the net stands with fetor chemicals (e.g. some phosphoroorganic pesticides) may help; endangered wader traps should not be active during the night time.

Domestic animals harm netted and trapped birds mainly when they are wild. Wild grown cats may pose a threat to netted birds, while wild dogs bring about more damage to trapped waders. **Remedy:** use of cat traps may solve the problem.

An electric fence may be helpful to exclude catching area from mammalian predators. However, sometimes fences should be very long, thus expensive. The other, less expensive solution could be ultrasonic devices used as mammals' repellents (dogs, martens and others).

14.3 Birds and the Weather/Habitat Factor

This is a parallel to the bird and the net/predator problem: birds in the nets and traps are exposed to weather and they may drown in rising water.

(a) A bird caught in a net is practically immobilized; its feathers pressed to the body, its head often pointing downwards, and its legs locked above the body. This unnatural position will affect the thermo-regulative abilities of the individual bird. At the same time, the whole body is exposed to heat, cold, rain, and insects more than usual. Low temperatures will cool the body below the physiologically acceptable level, and the bird will die if not removed soon enough. A similar threat is posed by high temperatures (above 30-32 °C) day or exposure to direct sunrays in sheltered place, e.g. in orientation test stand. The impact of the cold is aggravated if the bird gets wet. So, the first remedy to losses of this kind is regular and more frequent checks of the nets, particularly in rainy weather or when the fog condenses on the birds. Usually, short showers are harmless to the birds, but a heavy rain during a thunderstorm will kill small species. Birds removed from nets in rainy weather are wet and must be dried as soon as possible. In most cases storing them in dry cotton bag is enough, but note: synthetic fabric has a very low capacity to absorb water, and the bird in such bag should be dried with an artificial heat-source, but, caution must be taken not to overheat it! When the bird is soaked through and cold stiff, the best drying method is to put it directly against your breast, i.e. not in a bag or between shirt and sweater! The method is not pleasant to the bird holder, especially if the birds name is e.g. Woodpecker, but, remember that you are responsible for the situation! It is a good custom to change the control rhythm when a thunderstorm is approaching. E.g. many reedbed birds become very active and are caught more frequently just before the storm. So, weather losses may be partly reduced by good attendance to the nets.

Birds caught in mist-nets, particularly the ones hanging in the lowest shelf, also run the risk of being overlooked at a regular control walk. Under normal conditions, leaving a bird for one hour does not endanger its life. When the control is the last one in the evening, however, the bird will hang in the net for a couple of hours, and the following morning it will be dead. So, the evening control must be very scrupulous, using a good lamp and checking the whole length of each net. Shaking of the net up and down when lighted will reveal birds caught in the bottom shelf or near the top string (such sporadic catches are frequently overlooked). Cleaning the nets from leaves in the evening will facilitate the night controls.

(b) Birds netted in reedbeds, where water is standing under the nets, run the risk of drowning. At sites where the water level is stable, strict adherence to the advice in *Arrangement of the Netting Area: Wetland Habitats* will reduce accidental losses. In reedbeds, where the water level changes highly and frequently, netting is much more risky. First of all, the lowest string must be held much higher than in places with stable water; Water level should be continuously monitored, and when flood is expected, the nets should be pulled high.

Wader traps frequently get flooded by rising water and the birds run the risk of being drowned, particularly on sea-shores, and in lagoons subject to the changes of the overall sea level. The sea level changes with tides, currents, air pressure and direction and force of winds. In the daytime, the distribution of traps must be adjusted to changing water levels and weather. Any decision to leave active traps overnight should be based on knowledge of the local water situation; if there is any doubt, move the traps from the catching area to higher ground.

14.4 Removal and Transport of Birds Caught

The removal of birds from nets and traps is a potential source of loss and/or injury.

(a) When birds are freed from a net, it must be remembered that bird legs and bird wings (in spite of their flexibility) must not be moved too rapidly or with excessive power; never apply force perpendicular to a leg. This usually leads to a fracture; the most common injury when removing birds from nets. The same thing, injury to the bird, may happen if it is allowed frequently flapping of the wings. One single strong wing-beat may break the air sack connecting the body sack with the wing bone, or cause blood effusion to the lungs, and the bird will be unable to fly, at least for a couple of days. Most vulnerable to this kind of injury are juvenile bullfinches, chaffinches, greenfinches, flycatchers and tree pipits. Other species, like goldcrests, yellow wagtails and hirundines seem to be totally unaffected.

A great deal of bird mortality results from incorrect handling of birds during transport. **Remedy:** strictly apply all advice given within appropriate sections above (see p. 31). In spite of how inconvenient they may be to you, follow them!

14.5 Laboratory Work

Laboratory work is relatively safe to the birds. On peak days, apply the advice given in the *Alarm Routine* section. Remember that even under normal conditions, birds must never wait for ringing and/or additional processing while exposed to direct sun radiation. The ringing procedure frequently reveals all mistakes made by the staff when removing and transporting the birds: the ringer may find dead or injured birds in the bags, others are unable to fly after being ringed. Discuss the matter with the people responsible! Dead birds should be noted in the field-form and the cause of death, if known, should be given as a comment. This procedure does not help the dead bird, but it can serve to explain losses and find remedies against them.

Injured birds should be treated according to the particular kind of the injury:

(a) A bird with broken wing bones or leg fracture above the tarso-metatarsal joint should be put to death, since it has no chance for a normal life or at least must pay for life with prolonged pain.

(b) A bird's leg with an open fracture of the tarso-metatarsal joint or tarsus should be amputated ca. 5 mm below the joint (use sharp scissors). Birds with cut-off tarsi are observed rather frequently in the nature as victims of natural hazards, and they may be in quite good condition. Releasing such birds without amputation will cause prolonged pain and the broken leg may be fixed in an unnatural position, handicapping the individual for the rest of its life-time. However, it should be mentioned that amputation of the bird tarsus by laymen is illegal under welfare legislation of most EU countries, but transport of a small bird to the veterinary would be practically impossible and result in prolonged stress, pain and even death. Usually, for the small bird such amputation means one drop of blood, fixed quickly.

(c) A broken tarsus, where the bones are not displaced, should not be amputated since there is a chance that ends may join in a natural process; it is obvious that the bird should not be ringed on that leg and/or do not ring the bird at all.

(d) Birds that are unable to fly will be handicapped in different ways; some of them temporarily paralysed by fright, it is relatively common in thrushes; within a couple of minutes, they suddenly fly away showing no flight handicap whatsoever. Others cannot fly because of a broken air-sack or internal blood effusion, allow such birds to walk away on foot as there is no way to help them: some of them will die, others will be taken by raptors but some will recover and continue their migration (a Bullfinch with such an injury was recovered seven days later several hundred kilometres south of the ringing place).

Some wader species are especially vulnerable to stress after catching. This mainly applies to long-legged species such as the Bar-tailed Godwit, Curlew, Whimbrel and larger species of genus *Tringa*. One effect of stress may be a leg or wing cramp. Birds affected in this way look healthy but are unable to stand on their legs. Relatively few such cases occur when waders are caught with walk-in traps, more often when cannon or rocket nets are used, and most frequent with mist-nets, particularly if the birds are allowed to hang for some time. The possibility for stress myopathy also increases on hot days and in birds with poor condition. In order to minimize these effects, particular attention should be paid to the regularity of controls and to the proper construction of containers for birds. Vulnerable species and individuals that sit in the container instead of standing up should be ringed first. If a leg cramp occurs and the bird is unable to stand up and fly away, it should be left in peace away from people, preferably in a place where other individuals of the same species stay, for about 1-3 hours. If this does not help, the bird should be placed in a spacious and high (ca. 50 cm) container with free access of air and light, but not exposed to the sun! The bird should not be disturbed, the staff must avoid appearing in the bird's field of vision and to handle it any more. Water must be available in the container, but the edge of a rather deep vase containing water, must be placed at the ground level. If the leg cramp does not cease, it will be necessary to feed the bird. Smaller species quite willingly feed in captivity. Food should contain live "worms", e.g. earthworms, tubifex, nereids. The best way is to place them in a separate vase and supply 2-3 times a day. When the bird is unable to eat on its own it must be fed "by force". In addition a solution of glucose may be given every 2 hours. In most cases such a bird gets better within 5-6 days and can be freed. Some authors recommend giving ca. 0.5 mg, or in case of large birds, 1 mg of Valium (diazepam) with water and food just after the leg cramp has occurred. After such a treatment, the bird will sleep up to ten hours or even more. It should be left in peace in place that is not too warm. When it wakes up, it should be able to walk normally. Bathing the legs of birds with incipient cramp in water may also be an effective treatment in cold weather. In hot conditions it is advisable to reduce density of caught birds in containers by ringing a proportion of them (30-50%) from each compartment. Moreover, birds that are sitting in containers, and therefore are likely to become cramped, should be ringed and released first.

(e) A separate group of handicapped birds are those exhausted by the migration journey and then caught; they may lack energy reserves even to fly away from the laboratory. After some time, most of them will be able to move to nearby foraging areas and have a chance to rebuild their fat reserves, but some of them, regrettably, are unable to fly even this short distance and usually die. Such birds may survive if heated for a while or being given a glucose solution (one tea-spoon per hundred grams of water). Otherwise they will be victims of, a slightly artificial selection pressure eliminating the weakest individuals from the population. At any rate, try to minimize behaving as a selection tool!

If all remedies were applied, all possible cautions observed and care performed, and we have a dead bird, try to use its corpse for special studies. Sex criteria may be checked by inspecting the gonads, the lipid contents may be investigated, internal parasites collected (Figure 14.4), blood samples taken, etc. In that case the death of a bird will not only mean loss. But not all corpses are needed, and they should be instantly buried for sanitary reasons and to prevent bitter comments from the public.

Figure 14.4-1: Collecting special data: here tumours caused by trematodes. Wadi Allaqi, Egypt. Photo I. Rząd.

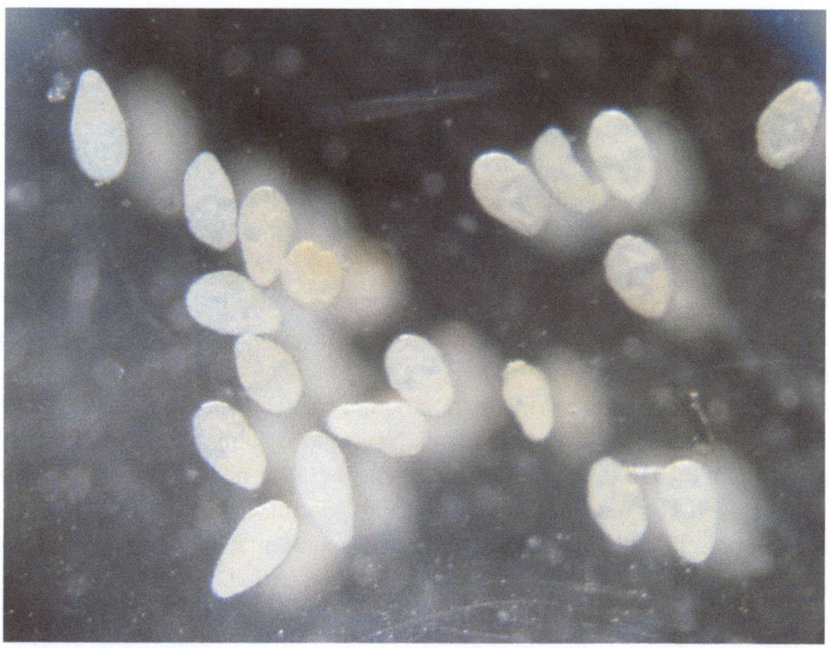

Figure 14.4-2: Collecting special data from a dead bird: high number of parasites in dead Swallow. Wadi Allaqi, Egypt. Photo I. Rząd.

15 Ringer's Safety and Health

Ringers' work at the ringing station is usually much safer than ringing of nestlings in nests located at high trees, exposed mountain rocks, or treacherous swamps. However, still there are some traps that can cause several dangerous situations, even danger of death, serious injuries or health problems. When working in real wilderness, pay attention to general safety, since the possibility of sudden flood or mud avalanche in a river valley exists, albeit rarely. In normal work, however, "normal" problems can occur. Some of such situations were already mentioned in the text above, but they should be repeated here:

15.1 General

Avoiding accidents. Clear the control path. You should be able to walk along it without any "gymnastics" in order to avoid branches, twigs, fallen trees, etc. When the ringer is in a hurry, his eyes may get hurt by a twig, leg broken over a branch, or accumulate other injuries by tripping and falling on rocks. Special care should be observed when fixing strings and stretching the nets; they cannot be put at two levels: at the level of your face, i.e. hitting the string when running can break eye glasses you may use, or injure your nose, and below your girdle; you may fall down, with all possible consequences, and you can even break your hand. For nocturnal net checks, use a good light source. Generally, everybody must be especially careful working at night, especially on water bodies and swamps (Figure 15.1).

Some birds in your hands could be directly dangerous to you. Hawfinches and shrikes may easily injure your hand by the strength of their bills, so the first thing to do when you remove such a bird; fix its head, and you can handle it partly with a bag on its head (Figure 15.2). Raptors and owls hit mainly with their claws, and they are very quick. Surprisingly, their hook-like bills are usually not as dangerous; however, there are some individual exceptions! The first thing when starting to remove a raptor or an owl; hold them by the tarsal joint of both legs; this is an exception to the usual suggestions when removing birds from nets. Owls may look as if they are sleeping, but the most dangerous moment is the initiation of the removal process. If, despite your care, the bird catches some part of you by its claws, remain calm and slowly turn the leg along the tail to the bird's back (remember Figure 5.14). The birds' claws will automatically open (because of an anatomical peculiarity of the leg). Jays use both methods of fight, bill and claws, and your response must be to use both tactics mentioned above. Tits are irritating with their pinching. Little bitterns are dangerous to your eyes when handled, for they have surprisingly long necks (Figure 15.3) and may hit your eye suddenly! The same applies to herons and bitterns. Additionally, be careful with moorhens and coots as well!

Figure 15.1-1: A dog guarding the line of nets. Aras, Eastern Turkey. Photo P. Busse.

Figure 15.1-2: Evening transport of walk-in traps – warning weather situation. Mouth of Reda, Poland. Photo W. Meissner.

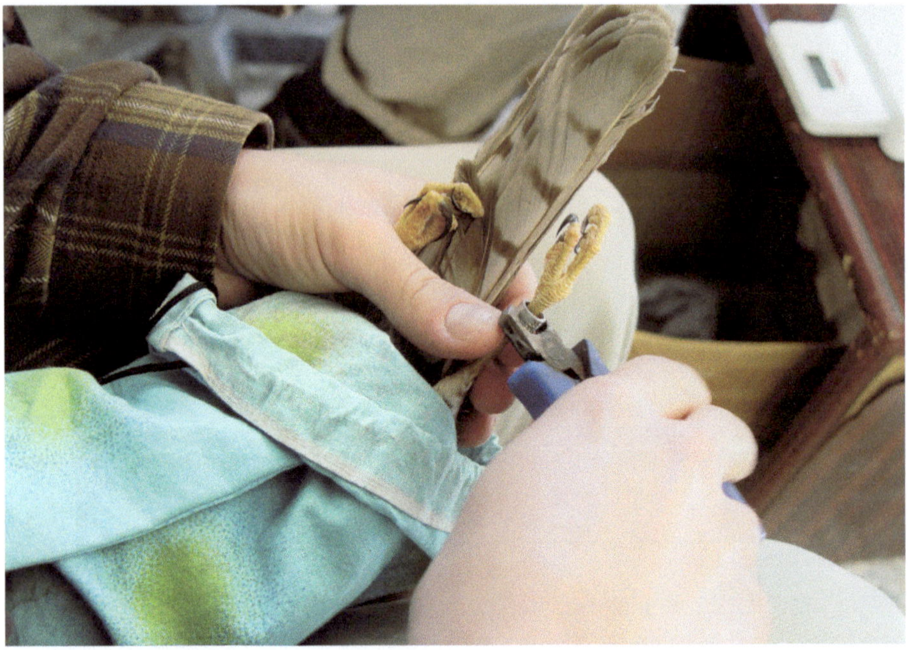

Figure 15.2-1: Safe ringing of the Sparrowhawk. Azraq, Jordan. Photo P. Busse.

Figure 15.2-2: Bigger birds should be released carefully. Azraq, Jordan. Photo P. Busse.

Figure 15.3: Little Bittern - „short" and „long" in a moment. Azraq, Jordan. Photo P. Busse.

15.2 Health Problems

Birds. Some special health problems can occur during ringers work with birds since they can be vectors of different, even potentially dangerous, diseases: avian flu and other virus, bacteria, or fungi originated illnesses. These are, fortunately, very uncommon in practice as most of ringers are careful enough with a general hygienic practices: washing hands, not using dirty bird bags for wiping noses or "cleaning" a kitchen table, etc. Anyway, remember to separate working and kitchen tables. The "bird table" must be absolutely free of sandwiches, cakes, chewing gums, cigarettes and drinks (only 70% alcohol for disinfection is allowed). There is no reason to panic about avian flu, but take care and have clean hands when eating. In any case, it is necessary to have a basic first aid kit with additional supplies for wound service, pain, allergy, cold, and stomach problems.

Additional health problems can be caused by animals other than birds that can be met during work at the station. These belong to very different systematic groups and can cause different problems at stations located in different regions.

Insects. They are the most common plagues at stations elsewhere; frequently, we encounter problems with mosquitos. In some regions, mosquitos make evening/night and sometimes early morning controls and work more difficult. However, in other areas, they can easily transfer really dangerous diseases, i.e. malaria, West Nile fever and other such deadly sicknesses. In regions where these hardest diseases do not exist, it is enough to use repellents on a skin and to protect the ringing stand with a mosquito net, e.g. a special hide with walls made of this kind of material (Figure 15.4). When ringing in a room or tent, some other repellents or fumigants can be useful. Working in malaria endemic regions requires use of special prophylactic medicines, while vaccinations are necessary against other health risks.

Flies can also be dangerous for health, since they easily transfer harmful microorganisms, especially in tropic areas but also in Europe. So, protect your food and the kitchenware against them. A special risk of a serious disease is connected with tsetse fly in Africa.

Ticks. In many regions of the world, ticks are problematic for people staying for an extended time in a field conditions. They are common vectors of dangerous viral and bacterial diseases, such as encephalitis or boreliosis. In some regions of Europe and Siberia, a high share of ticks is infected and dangerous. Against encephalitis, there is a vaccination available, but this must be done well in advance to the travel through this dangerous region. There is no vaccine against boreliosis, but it can be cured with antibiotics, especially when diagnosed early. Hence, care should be taken with ticks; check your body frequently, especially soft parts of a skin, since ticks like to place themselves in even the most intimate places. Check any itching places and remove ticks instantly using a pincette, not covering it by oil or twisting it as it is frequently advised. Be aware that a red, growing area around a tick can mean beginning of boreliosis, so immediately contact a doctor.

Figure 15.4: A hut with mosquito protecting walls for work where there is such problem.

Scorpions and spiders. These groups of animals could be dangerous in hot regions where bigger and more poisonous species live. They are active mostly at night, and some of them could hide in your shoes or clothes during the day. So, since clothes worn in the morning can be bitten or stitched by them, care must be taken to check all clothing and accessories before use. Frequently, scorpions are hidden under your tent, stones and pieces of wood you collect for fire. Usually, bites by smaller species or individuals are not mortal, but they are extremely painful.

Snakes. There are dangerous snakes as well as completely harmless ones. In Europe the only poisonous snake is the viper. There are two colour forms of this snake species: totally black and dark grey with pronounced black zig-zag stripe along the body. A bite of the viper is rarely mortal for an adult person, but it can be for a person with heart problems. There are available antitoxins, and in a case of a biting, contact a doctor immediately. In hot regions of the world, there are a lot of poisonous snakes and many of them are mortal, even with very short contact time, for an adult person. So, avoid of contact with unknown snakes, e.g. do not try to kill them, since there are cases where snakes can squirt poison from a distance, e.g. one of cobra species.

Mammals. This group of animals can be dangerous in two ways: big animals can kill a person directly in remote wilderness, e.g. Siberia (bears) or tropical regions (lions, tigers, elephants, buffaloes, rhinoceros or hippopotamus). Be careful, and obey directions of indigenous populations. In totally different ways, small mammals can be dangerous to your health, as they can infect you with rabies, which is a mortal illness if not treated quickly. For ringers, it is especially dangerous to remove bats that occasionally are caught in our nets, especially when we use those made of very thin thread. So, try to avoid being bitten by bat as much as possible during its removal. In European conditions, bigger species of bats need special care. Fruit eating bats of hot regions of the world are dangerous too, as they could transfer e.g. the Ebola fever, but they are rarely furious enough, like small bats, to bite humans.

15.3 People

The last, but not least, work collaboratively about local people if you work in certain areas of remote wilderness, but not only:

(a) In many places during migration of birds, intensive hunting is a common custom; the best is prophylactic so, inform local hunters that you are working there;

(b) In some European societies "green-fighters" could be dangerous to your nets; try to contact local activists and explain that you do not intend to kill birds;

(c) Unfortunately, in many places your equipment, e.g. nets, poles and even binoculars, could be stolen; especially, if you use expensive telescopic or good metal poles for nets, you leave loudspeakers for type-luring unattended or all persons leave the field camp simultaneously. If you are a dog lover and you have your dog in the station your belongings could be safer;

(d) In unstable regions, always have a local guard or at least local friends with you. This does not guaranty full safety, but it always assists in solving problems.

15.4 Life Conditions

Some words should be devoted to life conditions you encounter while staying and working at the station. There is a huge variety of living conditions you can find at a place. These are from living in normal house, with all comforts available, to tent camps in a real wilderness. The work could be done both in a laboratory with air conditioning and electric light at a table and at open places when temperatures approximate –10 °C or +35 °C. According to the situation, the logistics could be different, but some basic rules should be observed:

1. Organize sleeping and hygienic conditions to be as comfortable as possible – you must be fresh and full of energy to work,
2. Be equipped with raincoats and rubber boots of length adequate to the weather and areas you work in; this is especially important in cold weather and working

within wet biotopes; be ready to change wet clothes to dry ones, and to be able to dry them quickly,
3. In hot regions, do not forget sunglasses, hut and sun-protecting cream; mosquito repellents should be available as well; drink enough much and eventually supply loss of minerals with mineral tablets,
4. Organize as regular and as normal meals as possible; if you stay in the field for a long time, your body must work as usual and it cannot be stressed more than absolutely needed, and the condition of your stomach is a basis for your feeling and effectiveness in work.

You could not believe the above if you are young, but trust the experienced people; neglecting these rules will affect you when you are of certain age, and your ringing activity could be shortened by growing health problems. So, be careful and you will ring birds many, many years.

16 Alternative Methods of Holding and Measuring Birds

In some ringing manuals, other methods than what is described here as a standard methods of the bird holding, are described and shown at illustrations. Sometimes, they are recommended for general use, sometimes to perform separate measurements.

A left-hand handling is in common use as shown by figures in *The Ringer's Manual* (Spencer, 1972), Figure 16.1, or in *Identification Guide to European Passerines* (Svensson, 1992). The first method is very similar to that recommended here as the standard holding method (apart from the hand used for holding bird). The second one is completely different, as the bird is handled in the opposite direction, with the bill to the wrist. According to these methods of holding, standard descriptions of some measurements are adapted and given below.

Figure 16.1: Handling the bird – two methods after Spencer (1972) drawing.

Comments: The left-hand holding of bird derives from an old custom of ringers, even those working at the bird stations, that the same person handles the bird, rings and measures, and writes data into ringing form. When one uses this method, there is no necessity to move the bird from one hand to another when writing data: most people are right-handed and they write using their right hand "so, the right hand should be free of the bird". In a modern way of work, in a team, there is no such necessity anymore, and the speed of work can be much higher than before. As it was stressed in the description of the standard holding, the right-hand holding allows manipulation of the bird more quickly and safely to the bird since the right hand "feels" bird body much better.

Wing-length measurement. As the method of wing-length measurement used earlier, unflattened wing, is nearly not in use any more it will not be described here. It was rejected from the practice, as its results are not enough repeatable and possible to be standardised.

Wing-length flattened chord measurement after *The Ringer's Manual* (Spencer, 1972). This is basically the method described in *The Handbook of British Birds* (Witherby *et al.*, 1938-41): "The ruler is slipped under the naturally flexed (but unspread) wing and the carpal joint is pressed gently against the stop. The wing is then flattened against the rule by firm but gentle controlled pressure on the median or greater coverts. This removes some or all of the camber along and across the wing, but the primaries are not straightened, so that they lie along the rule with their normal lateral curvature undisturbed".

Comments. Although this method is capable of producing a reliable measurement of a wing, it should be noted that variation in the degree of pressure applied in holding the wing to the ruler would produce small differences in the measurements obtained. For this reason, the results obtained are less reproducible between ringers or, sometimes, even by the ringer himself. Nor is the method any more successful than method of unflattened wing at allowing unavoidable alteration of lateral curvature.

Wing-length maximum chord measurement after *The Ringer's Manual* (Spencer, 1972): "In this method, in addition to applying firm pressure on the wing as in method "flattened wing" to remove all camber along and across the wing, the lateral curvature is also eliminated as far as possible. This is done by sliding the wing forward along the rule until it meets the stop, straightening the bastard wing so that it falls into line, as far as possible, with the longest primary, and then straightening and extending the longest primary to its maximum length by stroking the thumb of the free hand along the shafts of primaries, from base to tip, pressing firmly against the rule all the while (Figure 16.2). It must be emphasized that no attempt must be made to pull the wing straight from the tip; a firm stroking action is required. Small differences in measurement may result from variation in the degree of straightness achieved, but the method reduces errors due to alteration of the lateral curvature during trapping and handling, or occasioned by dampness. It is, however, essential to keep the wing closed and parallel to the long axis of the bird's body.

Ruler with the stop is used. For birds small enough to be measured with a 30 cm ruler, it is recommended that the wing is measured to the nearest 1 mm".

Wing-length maximum chord measurement after *Manual of Field Methods* (Bairlein, 1995): "Wing length is determined as maximum chord which is the length of the flattened and straightened wing, and it is the distance between the bend of the wing and the longest primary.

Use the ruler with a stop at zero. Read to 0.5 mm.

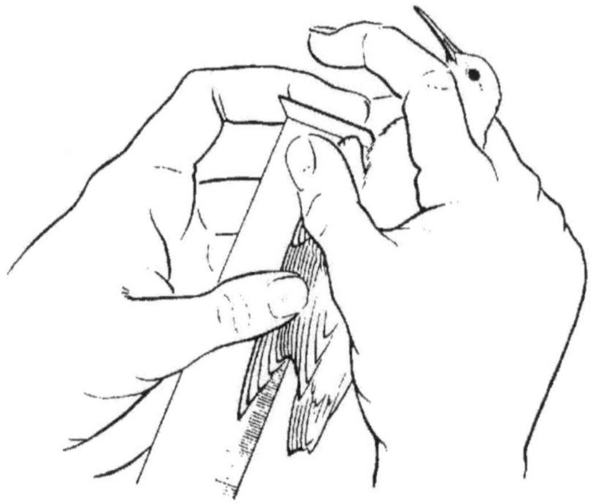

Figure 16.2: Wing-length measurement after Spencer (1972) drawing.

The wing should be folded ("resting position"; Figure 16.3). The wing is then flattened against the ruler with a gentle pressure on the primary coverts with a thumb. The primaries are straightened by pushing the thumb sideways (Figure 16.3B and C) until the primaries are parallel with the ruler. It is also of good help to adjust the position of the primaries with your index (2) or ring finger (3).

Straighten the wing, still flattened against the ruler by strokes with the thumb outwards along the shafts of primaries. Do not move the bend of the wing off the zero stop. Do not use excessive force, and be as cautious as possible to avoid any injures to the fragile wing bones and muscles".

Note that two methods of bird handling are applied in that description.

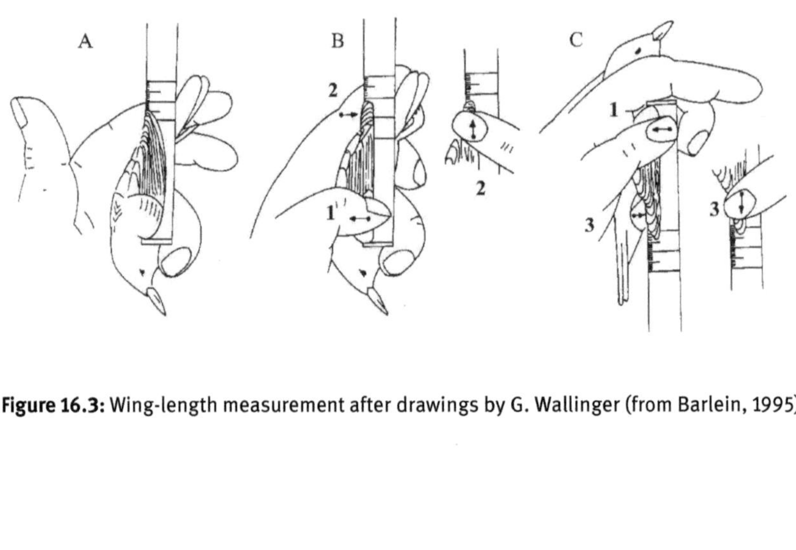

Figure 16.3: Wing-length measurement after drawings by G. Wallinger (from Barlein, 1995).

Comments. Differences in description of the maximum chord method as presented here, in relation to that recommended as the station standard are derived mainly from the other method of holding bird. The description derived from *The Ringer's Manual* (Spencer, 1972) is very close to that recommended in the standard description given in the main part of this manual: note the right-hand holding of the bird and position of index finger. Position of index finger here differs from that shown at Figure 16.3C illustrating the measurement given after *Manual of the Field Methods* (Bairlein, 1995). The position of index finger is important to exact feeling of location of the wing bend in relation to the ruler; the finger tip is most sensitive to feel pressure of the wing bend to the stop of the ruler. Exactness of control of the position of the bend in relation to the ruler is even higher when the ruler without a stop is used. There is a greater possibility to have longer readings when the ruler is placed as it is shown at Figure 16.3C, especially when one has tendency to much pull the primaries along the ruler. The left-hand holding of the bird does not influence the wing-length measurement if this is the only difference from the standard description or the description given in *The Ringer's Manual* (Spencer, 1972).

Holding the bird with the bill to the wrist, as shown at Figure 16.3 and in *Identification Guide to European Passerines* (Svensson, 1992), can be commented as in *The Ringer's Manual* (Spencer, 1972), "*... a method of measuring a wing with the bird held in the reverse grip. It is possible to measure equally accurately using this technique, but, it is potentially dangerous if a 30 cm ruler is used (because of difficulty of controlling both the wing and the heavy rule with one hand), there is nothing to recommend it*". In addition, it should be stressed that there is no other measurement, which could be done quickly and precisely enough when this method of the bird holding is used.

Wing-formula measurement. There are two other wing-formula measurement methods, than what is described in this manual, that formally intend to describe a wing-shape.

The method given in *Identification Guide to European Passerines* (Svensson, 1982): "When studying the wing-formula of a small passerine, it is often helpful to hold the bird in the left hand with the head towards the wrist and with the left wing very slightly spread between the right thumb and index finger. (...)

Make sure that the primaries generally forming the wing-point are not in moult, in which case they may not yet have their full length. Check both wings. When feathers are still growing, you will see generally the glossy, grey or greyish-white feather-sheaths at the base of the feathers (or gaps where feathers have been dropped) if the coverts are carefully lifted up with pliers, or by blowing on them. (The sheaths have a bluish tint during the first stages of growth). If one traps a bird with an unusually blunt wing, one should examine the bases of the outermost primaries by lifting up the under wing-coverts and look for remnants of the sheaths. Check also if any feather is accidentally lost or broken before studying the wing-formula. In museum specimens this will often be the case due to shots.

Gently put the tips of the feathers in order - they may become blunt while the bird is kept in a box or a bag. Make a note if the feathers are much abraded (edge of tips ragged). A heavily worn longest primary can easily be 3-5% shorter than its full length when fresh. To determine the position of the tip of the second primary, or a notch on the inner web of the second or third, spread the wing as little as possible. When you measure the distance between the tip of a primary or a secondary and the tip of the wing, use either dividers or a ruler (preferably transparent) placed to naturally folded wing, with the scale visible right against the tips (...). Do not measure the individual distances between the tips, a method which will be less accurate if the measurements are summed up. It is advisable to include the distance from the wing-point to the tip of the outermost secondary among these measurements".

Comments. This method is basically the same as our standard wing-formula measurement; it comprises distances between tips of subsequent primaries from the wing-tip. However, the method of the bird handling and measurement technique described do not allow its use for wing-formula differentiation studies on live birds, as it is very inconvenient and extremely time consuming; trials of applying it into the station routine lead consequently to rejecting this very important measurement from the station practice. In practice, it could be used for single individuals when the wing-formula measurement is needed in the identification process, and the method description suggests such use of it. It should be stressed that the comments on checking whether the primaries are not growing are of great importance for any wing-formula measurement.

The method given in *Manual of Field Methods* (Bairlein, 1995, after Jenni & Winkler, 1989) and called "wing-shape" measurement has really close to nothing meaning for description of the real wing-shape of the wing treated as a functional unit; it contains several measurements of the length of subsequent flight feathers. Therefore, the description of this measurement is given in the *Additional Measurements and Scores* section (p. 97).

Tail-length measurement. There are a few methods of tail-length measuring taking the tail-length from the base of rectrices to the tip of the longest one.

The method given in *Identification Guide to European Passerines* (Svensson, 1982), measurement taken from below the tail: „Start with a moult examination and put the tips of the feathers in order. Preferably use a thin ruler with the scale starting from the outer (very thin or pointed) edge of one end. Place that end under the tail between the tail-feathers and the under tail-coverts and push it gently against the root of the central pair of tail-feathers (...). Measure to tip of the longest tail-feather when the tail is naturally folded".

Another version of the method given in *The Ringer's Manual* (Spencer, 1972): "Alternatively, dividers may be used, as shown at Figure 16.4. Hold the dividers in the same plane as the tail so that is the side of the point which impinges against the body".

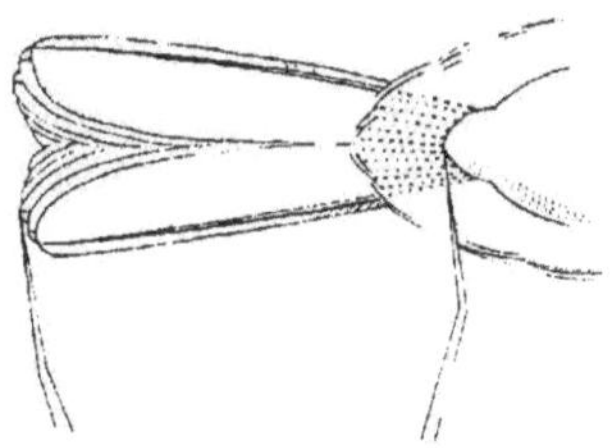

Figure 16.4: Tail-length measurement using dividers after Spencer (1972) drawing.

The above described variants of the measurement taken from the below the tail have their counterparts in such measurements taken from the above of the tail - between the tail-feathers and the upper tail-coverts.

Comments. All these methods give different results not only if one compares themselves but also in comparison with the standard method "to the back" described in this manual – so, the method of measurement must be stated. These methods could be a little bit risky to the bird, especially those which manipulate with pointed tools close to the preen gland. The classic method (for skin studies) as described in *The Handbook of British Birds* (Witherby *et al.*, 1938-41) – use dividers perpendicularly to the tail surface with one divider leg located between central rectrices cannot be used for alive birds as the risk of damage to the preen gland or pygostyl is high.

Fat-scoring. The alternative method of the fat-scoring is recommended in *Manual of Field Methods* (Bairlein, 1995 after Kaiser, 1993): "The size of the visible fat depot is determined with the use of a 9-grade score (0-8) – Figure 16.5.

Two of the most important fat deposits are checked, the furcular (intraclavicular depression, "tracheal pit") and the abdominal. A specific positioning of the bird's body is necessary to make the determinations.

The bird is laid on its back in one hand, and the legs are held by the other hand. The neck must be stretched slightly so that the furcular deposit is well visible, and the feathers must be blown aside. Legs of the bird should be spread aside, not pulled up or down – it will move the fat.

Additional requirements are (1) the use of bright light, which intensifies the contrast between yellowish fat layers and red muscle tissue, and (2) the determination of the amount of visible fat before the bird is weighed to avoid biasing the measuring process.

The scores from 0 to 8 are taken in the following manner using the subclass description (Table 16.1). At first, estimate the fat class at the furcular region. For example, if the fulcular is "filled", i.e. not concave or convex bulging, the subclass corresponds to 4.00, 4.25 or 4.50.

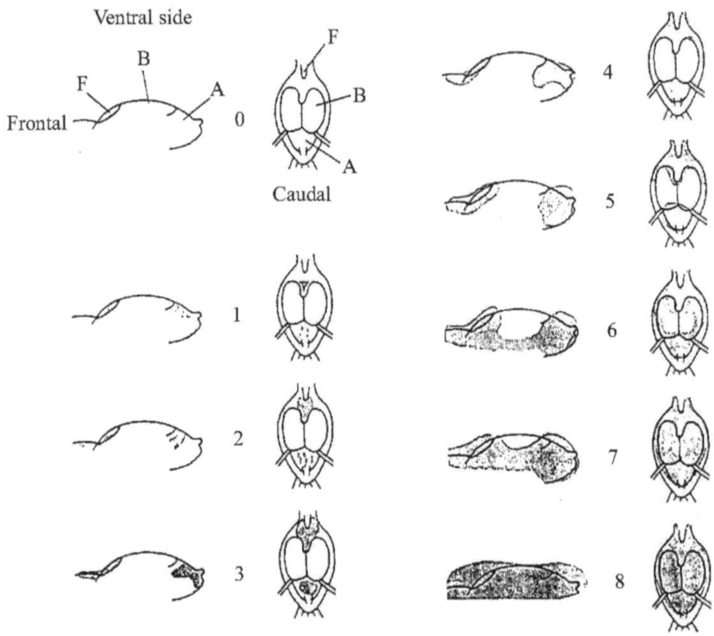

Figure 16.5: Fat scoring after Kaiser, 1993 (from Bairlein, 1995) – compare Table 6.3.

Secondly, check the abdominal area. If the fat deposit covers abdominal structures completely, and the liver is not visible, but the abdominal fat layer is not convex bulging, total score is 4.

However, some individuals do not follow the process in fat deposition as shown in Figure 16.5. For example, the abdominal area may have a slightly rounded pad of fat with intestinal loops not visible (3.50), while the furcular depression is still not completely covered with fat (1.75). Thus, the average amount of fat score is. 2.6, recorded as main fat score 3.

Use only main fat classes 0 to 8".

Comments. The method is based on the same idea as the standard description in this manual. However, description of the subclasses is much more complicated to apply. From the observation of trained people, using the method of averaging the subclasses results in the final main classes being purely theoretical, especially in a hurry, and the results are based on personal judgement depending on one fat deposit only. It was checked because of psychological reasons, one person could believe more the furcular depot while other is more convinced with the abdomenal one. So, the results are less comparable, although derived from the procedure, which seems apparently more precise.

Table 16.1: Description of the fat classes (from Kaiser, 1993, after Bairlein, 1995).

Main class	Subclass	Furcular depression	Abdomen	Colour of the considered areas
0	0.00	No fat	No fat	Dark red
	0.25	Barest trace, very narrow stripe	Fat deposits not yet delimited	
	0.50	Small stripe	As above	Red
	0.75	Wedge-shaped	Small trace, patchy	Light red
1	1.00	Wide wedge	Trace, very small stripes around intestinal loops (mm)	Light red
	1.25	Half of fulcular depression is covered	Trace, stripes 1 mm wide	Yellow-red
	1.50	Almost completely covered with fat	Trace, stripes smaller than intestinal loops	Yellow-red
	1.75	Small amount, almost completely covered with fat	Wide stripes (2 mm)	Yellowish
2	2.00	Completely covered, shape deeply concave	Slips of visceral fat, area between intestinal loops completely filled	Light yellow
	2.25	Completely covered, shape deeply concave	Some subcutaneous lipid, not yet forming pad	Light yellow
	2.50	Completely covered, shape deeply concave	Very small pad	Light yellow
	2.75	Completely covered, shape deeply concave	Small pad, at least 2 or 3 intestinal loops still visible	Light yellow
3	3.00	Moderate fat reserves cover ends of interclavicles	Flat pad, one loop still visible	Light yellow
	3.25	Concave	Slightly rounded pad, one loop sometimes visible	Yolk-yellow
	3.50	Still concave	Slightly bulging, loops completely covered	Yolk-yellow
	3.75	Almost filled	Bulging, liver visible	Yolk-yellow
4	4.00	Filled up to distal portion of interclavicles	Conspicuously bulging (2-4 mm), liver sometimes visible	Yolk-yellow
	4.25	Filled up to distal portion of interclavicles	Further increase in bulge (4-5 mm), liver sometimes visible	Yolk-yellow
	4.50	Filled up to distal portion of interclavicles	Abdominal structures completely covered, liver not visible	Yolk-yellow

continued **Table 16.1:** Description of the fat classes (from Kaiser, 1993, after Bairlein, 1995).

Main class	Subclass	Furcular depression	Abdomen	Colour of the considered areas
	4.75	Slightly bulging with central depression (concave)	Abdominal structures completely covered and bulging	Yolk-yellow
5	5.00	Convex bulge	Extreme convex bulge, increasing thickness	Yolk-yellow
	5.25	Just covering flight muscles from either furc., or abdomen	Extreme convex bulge, increasing thickness	Yolk-yellow
	5.50	Covering border of flight muscles a few mm	Covering border of flight muscles a few mm	Yolk-yellow
6	6.00	Covering flight muscles by several mm	Covering flight muscles by several mm	Yolk-yellow
	6.50	Fat reaches flight muscles from sides of wings		Yolk-yellow
	6.75	Fat covering flight muscles conspicuously		Yolk-yellow
7	7.00	Three quarters of flight muscles covered		Yolk-yellow
	7.25	Large rounded fat-free area in middle of breast		Yolk-yellow
	7.50	Small rounded fat-free area (red)		Yolk-yellow
	7.75	Very small fat-free area still visible		Yolk-yellow
8	8.00	Flight muscles not visible, fat layer covers underside/ ventral side of the bird completely		Yolk-yellow

17 Non Standard Ringing Procedures

The manual is intended to deal mainly with ringing as a tool to study bird migration. However, the bird ringing as a method could be used in different ecological research. This can be run as a part of the basic work of the permanent bird station or as programmes for ordinary ringers participating in them as volunteers. The most known are projects following population at breeding grounds: colour bird ringing and Constant Effort Sites programme. At some areas where a lot of birds spend a winter ringing projects at bird feeders are popular. Because some of these activities could be effectively run at permanent ringing stations, a bit of information about could be justified in the manual, especially that there can be used procedures similar to traditional bird ringing.

Colour ringing. Tagging of birds using colour markers – colour rings, wing tags, neck collars – is very close to the classic ringing. Generally, colour rings are similar to normal metal rings, but made of colour plastic, but differentiated by colour, not by inscriptions and numbering (although some of them are bearing numbers). After colour ringing, the bird or the bird group is identified from a distance by colour or combination of colour rings; it is not necessary to re-trap the bird to have return information on the bird. This makes collecting ecological information about birds at breeding area much more efficient than waiting for subsequent catches. Colours give possibility to know at least the group the individual bird belongs to: the defined cohort is marked by colour, e.g. nestlings from the year 2010 bear yellow ring, from 2011: red ring, while those from 2012: blue ring, and during breeding season 2013 we will know the exact age of birds breeding in the vicinity. The more sophisticated colour ringing uses combination of several such rings put on one or two legs, right and left: so, yellow ring on the left leg does not mean the same as the yellow ring on the right one. If we use a few rings possibility of individual coding grows and return information, obtained without re-trapping bird grows much. It must be mentioned that colour ringing is the only case when putting more than one ring on one leg is allowed. The disadvantage of common colour rings is that plastic is much less durable than metal and birds can lost some colour rings relatively soon. This problem is solved in the European Laser Signed Advanced ring rings system (ELSA) designed for White Stork ringing and intended that the ring number to be read by binoculars or telescope rather than waiting for traditional bird recovery report. By the way, the idea of reading ring number from a distance is nowadays more and more popular as optical equipment is more accessible and modern cameras with high resolution bring readable pictures easily.

The similar idea of identification from a distance is used in a form of wing tags and neck collars; the readings can be easily read, even using common binoculars.

Radio tagging. Quick development of electronic devices brought important progress in bird migration studies. At the beginning, there were small radio transmitters, that were fixed on a bird, allowed locating it using directional antenna from a distance hundreds meters to a few kilometres. In the bird migration study, they were useful for studying stopover behaviour locally. This kind of radio transmitters still could be useful in such research, as these devices are relatively cheap. However, the work with them is tiring and time consuming. Recent development of miniaturized tracking technology has opened a new perspective in bird migration studies. They differ in applied tracking technology, accuracy and weight (Table 17.1, Figure 17.1). Tracking device deployed on a given bird species must be small enough to be carried without difficulty. It is widely accepted that weight of the tracking device should not exceed 3% of body mass of the bird. Nowadays only geolocator technology may be applied for small birds, weighing not less than 17-20 g. Geolocator mounted on the lower back or on the upper part of the leg of a bird records light levels in relation to an internal timer. According to these light data the latitude and the longitude could be estimated by specialized computer program. Accuracy of geolocators is severely limited in equatorial areas during some parts of the year and in Polar Regions. However, geolocator accuracy is low and varies depending on location, habitat and weather, they are sufficient in studies on long distance migrants (e.g. Egevang *et al.*, 2010; Bairlein *et al.*, 2012; Cormier *et al.*, 2013).

Table 17.1: General characteristics of the most popular bird tracking devices in 2013.

Tracking technology	Minimum mass of the device [g]	Accuracy	Advantages	Disadvantages
GPS with satellite relay systems	20		High accuracy. Global range of operation; SOLAR	Expensive; not more than 20 locations per day
GPS logger with radio data transmitter (UHF)	5 g for short (about 400 m) range of transmission; 15 g for long (about 4 km) range of transmission	2-5 m	High accuracy; SOLAR	Data downloading possible only when birds stay within the range of antenna
GPS logger with GSM transmission	15	2-5 m	High accuracy; data transmission through cell phone networks; SOLAR	Only for larger or medium sized birds
Solar geolocation	0.5	50–200 km	Cheap and light; might be used for tracking small passerines (about 15 g of body mass) data	Low accuracy; require recovery of the device to download data

Figure 17.1-1: Mallard with GPS device. Photo W. Meissner.

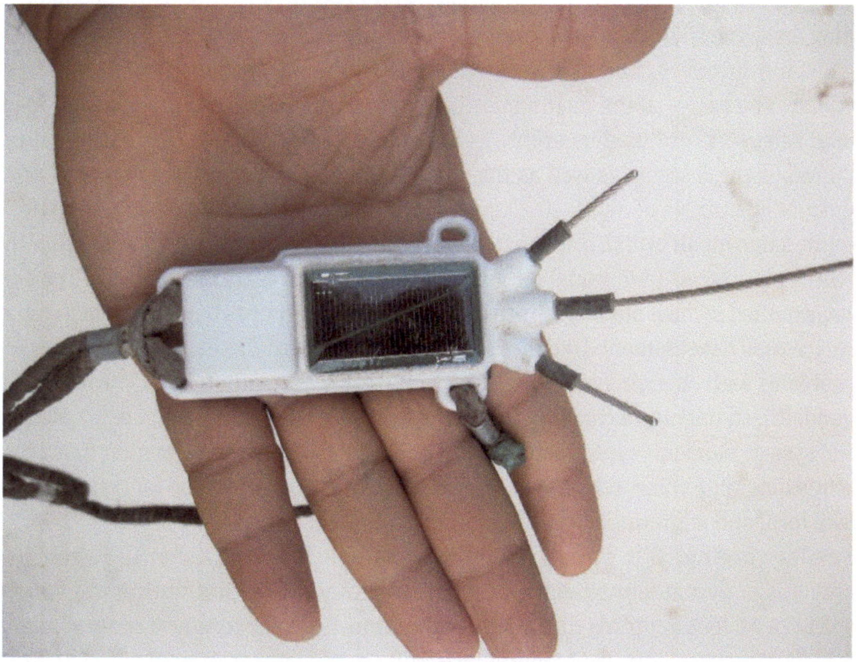

Figure 17.1-2: GPS device from the White Stork found near Aswan, Egypt. Photo I. Rząd.

Devices based on GPS systems usually record and store location data at a predetermined interval. GSM module allows changes of operating parameters remotely in loggers set on animals. Data stored in logger memory (usually up to 30 000 GPS positions) are relayed to a central data store or internet-connected computer using an embedded cellular (GPRS), radio, or satellite modem. These devices offer high accuracy of obtained geographical positions of tracked birds, hence, they may be used both for studies on local bird movement and studies on long distance migration. It can be expected that in the future, more accurate devices will be available for tracking small birds, geolocators are currently the only options for majority of passerines and small sized waders.

There is no doubt that we are entering a new era in bird migration studies. Nowadays, information on migratory routes may be obtained by combining data from bird rings returns, analyses of biomarkers (e.g., genes and stable isotopes), cage orientation tests and applying modern tracking devices. However, classic studies based on bird ringing and measuring are still widely used providing valuable data not only on migration, but also on mortality, longevity, site fidelity and many other aspect of birds life.

Constant Effort Site. This programme, working in several European countries, is generally set for voluntary ringers, but there is no reason to not apply it in a permanent ringing station. The basic idea is to catch the birds in a standardized manner during breeding period. Therefore, the idea is very close to that recommended in this book for the bird migration studies. CES programme monitors numbers of birds inhabiting the defined breeding area, gives information about productivity, from a proportion between number of adults and juveniles, and estimates adult survival rates, for species with breeding site fidelity. The basics of the method used are: twelve ringing visits at the site between May and August, distributed as evenly as possible. The standard set of nets should be used throughout all visits, as well as the nets should be located in the same positions. Recommended number of nets is 10-20 ones, and a time of work should be not less than 6 hours (the same for all visits), starting from the dawn. The scheme allows some flexibility as to these parameters (additional nets as well as additional visits are allowed). The site habitat is limited to more stable wet, scrub and deciduous biotopes being relatively stable as to succession development. Habitat should be carefully described at the start of the site. The problems with changing habitat are the same as it was mentioned for sites used for the standardized migration catching (see *Arrangement of the Netting Area* - p. 51). Within the CES system biometrics information is not required, but the ringing station that work in the breeding time will surely collect these data, because of their value for comparisons between local and migrating populations.

Feeder ringing. Out of migrations seasons and breeding time (CES) ringers, and obviously, permanent ringing stations frequently do ringing during the winter (Figure 17.2). As feeding birds during the winter time is a very common custom of not only ornithologists, ringing at feeders is easy and could be very effective; many birds are attracted by food supplied to the feeder (see also *Attracting the Birds to Nets and*

Traps – p. 75). Ringing at feeder should be carried out with a special care about safety of birds, among others it must be avoided too much disturbance to the birds when weather conditions are really critical – very low temperature, deep snow cover and, especially, when trees are covered by a slimy ice cover or when cold fog make the birds wet quickly. It is recommended to extract caught birds frequently, even just after they are caught. You need to remember that as the feeder, you are carefully controlled by

Figure 17.2-1: The birds caught at a feeder in winter. Przebendowo, Poland. Photo P. Busse.

Figure 17.2-2: A flock of birds at the feeder in winter. Przebendowo, Poland. Photo P. Busse.

18 Julian Date and Pentade Numbering

In many cases Julian Date (number of a day within a year) and five day periods numbering are useful in data reporting, presenting and analyses. These are shown in the Table 18.1.

Table 18.1: Julian dates and pentade numbers.

1	**1**	1 Jan.	31	**7**	31 Jan.	61	**13**	2 Mar.	91	**19**	1 Apr.
2		2 Jan.	32		1 Feb.	62		3 Mar.	92		2 Apr.
3		3 Jan.	33		2 Feb.	63		4 Mar.	93		3 Apr.
4		4 Jan.	34		3 Feb.	64		5 Mar.	94		4 Apr.
5		5 Jan.	35		4 Feb.	65		6 Mar.	95		5 Apr.
6	**2**	6 Jan.	36	**8**	5 Feb.	66	**14**	7 Mar.	96	**20**	6 Apr.
7		7 Jan.	37		6 Feb.	67		8 Mar.	97		7 Apr.
8		8 Jan.	38		7 Feb.	68		9 Mar.	98		8 Apr.
9		9 Jan.	39		8 Feb.	69		10 Mar.	99		9 Apr.
10		10 Jan.	40		9 Feb.	70		11 Mar.	100		10 Apr.
11	**3**	11 Jan.	41	**9**	10 Feb.	71	**15**	12 Mar.	101	**21**	11 Apr.
12		12 Jan.	42		11 Feb.	72		13 Mar.	102		12 Apr.
13		13 Jan.	43		12 Feb.	73		14 Mar.	103		13 Apr.
14		14 Jan.	44		13 Feb.	74		15 Mar.	104		14 Apr.
15		15 Jan.	45		14 Feb.	75		16 Mar.	105		15 Apr.
16	**4**	16 Jan.	46	**10**	15 Feb.	76	**16**	17 Mar.	106	**22**	16 Apr.
17		17 Jan.	47		16 Feb.	77		18 Mar.	107		17 Apr.
18		18 Jan.	48		17 Feb.	78		19 Mar.	108		18 Apr.
19		19 Jan.	49		18 Feb.	79		20 Mar.	109		19 Apr.
20		20 Jan.	50		19 Feb.	80		21 Mar.	110		20 Apr.
21	**5**	21 Jan.	51	**11**	20 Feb.	81	**17**	22 Mar.	111	**23**	21 Apr.
22		22 Jan.	52		21 Feb.	82		23 Mar.	112		22 Apr.
23		23 Jan.	53		22 Feb.	83		24 Mar.	113		23 Apr.
24		24 Jan.	54		23 Feb.	84		25 Mar.	114		24 Apr.
25		25 Jan.	55		24 Feb.	85		26 Mar.	115		25 Apr.
26	**6**	26 Jan.	56	**12**	25 Feb.	86	**18**	27 Mar.	116	**24**	26 Apr.
27		27 Jan.	57		26 Feb.	87		28 Mar.	117		27 Apr.
28		28 Jan.	58		27 Feb.	88		29 Mar.	118		28 Apr.
29		29 Jan.	59		28 Feb.	89		30 Mar.	119		29 Apr.
30		30 Jan.	60		1 Mar.	90		31 Mar.	120		30 Apr.

continued **Table 18.1:** Julian dates and pentade numbers.

121	**25**	1 May	151	**31**	31 May	181	**37**	30 Jun.	211	**43**	30 Jul.
122		2 May	152		1 Jun.	182		1 Jul.	212		31 Jul.
123		3 May	153		2 Jun.	183		2 Jul.	213		1 Aug.
124		4 May	154		3 Jun.	184		3 Jul.	214		2 Aug.
125		5 May	155		4 Jun.	185		4 Jul.	215		3 Aug.
126	**26**	6 May	156	**32**	5 Jun.	186	**38**	5 Jul.	216	**44**	4 Aug.
127		7 May	157		6 Jun.	187		6 Jul.	217		5 Aug.
128		8 May	158		7 Jun.	188		7 Jul.	218		6 Aug.
129		9 May	159		8 Jun.	189		8 Jul.	219		7 Aug.
130		10 May	160		9 Jun.	190		9 Jul.	220		8 Aug.
131	**27**	11 May	161	**33**	10 Jun.	191	**39**	10 Jul.	221	**45**	9 Aug.
132		12 May	162		11 Jun.	192		11 Jul.	222		10 Aug.
133		13 May	163		12 Jun.	193		12 Jul.	223		11 Aug.
134		14 May	164		13 Jun.	194		13 Jul.	224		12 Aug.
135		15 May	165		14 Jun.	195		14 Jul.	225		13 Aug.
136	**28**	16 May	166	**34**	15 Jun.	196	**40**	15 Jul.	226	**46**	14 Aug.
137		17 May	167		16 Jun.	197		16 Jul.	227		15 Aug.
138		18 May	168		17 Jun.	198		17 Jul.	228		16 Aug.
139		19 May	169		18 Jun.	199		18 Jul.	229		17 Aug.
140		20 May	170		19 Jun.	200		19 Jul.	230		18 Aug.
141	**29**	21 May	171	**35**	20 Jun.	201	**41**	20 Jul.	231	**47**	19 Aug.
142		22 May	172		21 Jun.	202		21 Jul.	232		20 Aug.
143		23 May	173		22 Jun.	203		22 Jul.	233		21 Aug.
144		24 May	174		23 Jun.	204		23 Jul.	234		22 Aug.
145		25 May	175		24 Jun.	205		24 Jul.	235		23 Aug.
146	**30**	26 May	176	**36**	25 Jun.	206	**42**	25 Jul.	236	**48**	24 Aug.
147		27 May	177		26 Jun.	207		26 Jul.	237		25 Aug.
148		28 May	178		27 Jun.	208		27 Jul.	238		26 Aug.
149		29 May	179		28 Jun.	209		28 Jul.	239		27 Aug.
150		30 May	180		29 Jun.	210		29 Jul.	240		28 Aug.

continued **Table 18.1:** Julian dates and pentade numbers.

241	**49**	29 Aug.
242		30 Aug.
243		31 Aug.
244		1 Sep.
245		2 Sep.
246	**50**	3 Sep.
247		4 Sep.
248		5 Sep.
249		6 Sep.
250		7 Sep.
251	**51**	8 Sep.
252		9 Sep.
253		10 Sep.
254		11 Sep.
255		12 Sep.
256	**52**	13 Sep.
257		14 Sep.
258		15 Sep.
259		16 Sep.
260		17 Sep.
261	**53**	18 Sep.
262		19 Sep.
263		20 Sep.
264		21 Sep.
265		22 Sep.
266	**54**	23 Sep.
267		24 Sep.
268		25 Sep.
269		26 Sep.
270		27 Sep.

271	**55**	28 Sep.
272		29 Sep.
273		30 Sep.
274		1 Oct.
275		2 Oct.
276	**56**	3 Oct.
277		4 Oct.
278		5 Oct.
279		6 Oct.
280		7 Oct.
281	**57**	8 Oct.
282		9 Oct.
283		10 Oct.
284		11 Oct.
285		12 Oct.
286	**58**	13 Oct.
287		14 Oct.
288		15 Oct.
289		16 Oct.
290		17 Oct.
291	**59**	18 Oct.
292		19 Oct.
293		20 Oct.
294		21 Oct.
295		22 Oct.
296	**60**	23 Oct.
297		24 Oct.
298		25 Oct.
299		26 Oct.
300		27 Oct.

301	**61**	28 Oct.
302		29 Oct.
303		30 Oct.
304		31 Oct.
305		1 Nov.
306	**62**	2 Nov.
307		3 Nov.
308		4 Nov.
309		5 Nov.
310		6 Nov.
311	**63**	7 Nov.
312		8 Nov.
313		9 Nov.
314		10 Nov.
315		11 Nov.
316	**64**	12 Nov.
317		13 Nov.
318		14 Nov.
319		15 Nov.
320		16 Nov.
321	**65**	17 Nov.
322		18 Nov.
323		19 Nov.
324		20 Nov.
325		21 Nov.
326	**66**	22 Nov.
327		23 Nov.
328		24 Nov.
329		25 Nov.
330		26 Nov.

331	**67**	27 Nov.
332		28 Nov.
333		29 Nov.
334		30 Nov.
335		1 Dec.
336	**68**	2 Dec.
337		3 Dec.
338		4 Dec.
339		5 Dec.
340		6 Dec.
341	**69**	7 Dec.
342		8 Dec.
343		9 Dec.
344		10 Dec.
345		11 Dec.
346	**70**	12 Dec.
347		13 Dec.
348		14 Dec.
349		15 Dec.
350		16 Dec.
351	**71**	17 Dec.
352		18 Dec.
353		19 Dec.
354		20 Dec.
355		21 Dec.
356	**72**	22 Dec.
357		23 Dec.
358		24 Dec.
359		25 Dec.
360		26 Dec.
361	**73**	27 Dec.
362		28 Dec.
363		29 Dec.
364		30 Dec.
365		31 Dec.

References

Publications from where at least one exact citation of a text or an illustration is included ("source publications") are given in bold.

Bainbridge, I. (1976). Curlew, cramp and keeping cages. Wader Study Group Bulletin, 16, 6-8.

Bairlein, F. (1995). Manual of Field Methods. Wilhelmshaven, Germany: European-African Songbird Migration Network.

Bairlein F., Norris D.R., Nagel R., Bulte M., Voight C.C., Fox J.W., Hussell D.J.T., Schmaljohann H. (2012). Cross-hemisphere migration of a 25 g songbird. Biology Letters, 8 (5), 505–507.

Belopolsky, L.O., Bekzhanova, D.S., Mezhenny, A.A., & Erik, V.V. (1959). On the study of bird migration by means of big traps. 2nd All-Union Ornithological Conference Abstracts, 2, 105-107. (in Russian)

Bub, H. (1991). Bird trapping and bird banding: a handbook for trapping methods all over the world. Ithaca, NY: Cornell University Press.

Busse, P. (1974). Biometrical methods. Notatki Ornitologiczne, 15, 114-126. (in Polish)

Busse, P. (1983). Biometrical standards in the Operation Baltic work. Ring, 116, 125-138.

Busse, P. (1986). Wing-shape indices and the problems with their interpretation. Notatki Ornitologiczne, 27, 139-155.

Busse, P. (1995). New technique of a field study of directional preferences of night passerine migrants. Ring, 17, 97-116.

Busse, P. (2000). Bird Station Manual. Gdańsk: University of Gdańsk Press.

Clark, N.A. (1986). Keeping-cages and keeping-boxes. Wader Study Group Bulletin, 46, 32-33.

Clark, N.A., & Austin, G.E. (2005). The use of tape recordings of roosting wader flocks to increase wader mist netting success. Wader Study Group Bulletin, 107, 46-49.

Cormier R.L., Humple D.L., Gardali T., Seavy N.E. (2013). Light-level geologgers reveal strong migratory connectivity and within-winter movements for a coastal California Swainson's Thrush (*Catharus ustulatus*) population. Auk, 130, 283–290.

Dolnik, V. R., & Payevsky, V. A. (1976). Rybachy-type trap. In: Ringing in the study of bird migrations in the USSR (pp. 73-81). Moscow: Nauka Press. (in Russian)

Egevang C., Stenhouse I.J., Phillips R.A., Petersen A., Fox J.W., Silk J.R.D. (2010). Tracking of Arctic Terns *Sterna paradisaea* reveals longest animal migration. Proceedings of the National Academy of Sciences USA, 107, 2078–2081.

Erik W.W. (1967). The big trap for mass bird-trapping. In: R.L. Potapov, B.E. Vykhovski (Eds), Migratsii ptits Pribaltik (Cisbaltic bird migration) (pp. 51-55). Leningrad: "Nauka" Publishing House. (in Russian)

Figuerola, J., & Gustamante, L. (1995). Does use of a tape lure bias samples of Curlew Sandpipers captured with mist nets? Journal of Field Ornithology, 66, 497-500.

Gavrilov, E. I. (1968). New model of the catching box - Heligoland trap. Ring, 56, 143-144.

Gosler, A.G., Greenwood, J.J.D., Baker, J.K., & King J.R. (1995). A comparison of wing length and primary length as size measures for small passerines. A report to the British Ringing Committee. Ringing and Migration, 16, 65-78.

Green, G. H. (1980). Total head length. Wader Study Group Bulletin, 29, 18.

Gregory, N. C., & Robins, J. K. (1998). A body condition scoring system for layer hens. New Zealand Journal of Agricultural Research, 41, 555-559.

Harrington, B. (2004). Use care in determining age-ratios in shorebirds: they may differ relative to flock position, flock location and behaviour. Wader Study Group Bulletin, 104, 92–93.

Haase, B. (2002). The use of tape-recorded distress calls to increase shorebird capture rates. Wader Study Group Bulletin, 99, 58–59.

Jenni, L., & Winkler, R. (1989). The feather-length of small passerines: A measurement for wing-length in live birds and museum skins. Bird Study, 36, 1-15.

Jezerskas L. (1990). A new-type trap "Zigzag" for catching birds in Ventes Ragas. Acta Ornithologica Lithuanica, 2, 157-165.

Kaiser, A. (1993). A new multcategory classification of subcutaneous fat deposits of songbirds. Journal of Field Ornithology, 64, 246-255.

Lessels, C. M., & Leslie, R. (1977). Alternative wader catching. Wader Study Group Bulletin, 20, 17–21.

Lindström, Å., Klaassen, M., & Lanctot, R. (2005). The foldable "Ottenby" walk-in trap: a handy and efficient wader trap for expedition conditions. Wader Study Group Bulletin, 107, 50-53.

Mehl, K.R., Drake, K.L., Page, G.W., Sanzenbacher, P.M., Haig, S.M., & Thompson, J.E. (2003). Capture of breeding and wintering shorebirds with leg-hold noose-mats. Journal of Field Ornithology, 74, 401–405.

Meissner, W. (1992). Death of waders at ringing points of WRG "KULING" at Reda mouth and Rewa. Ring, 14, 109-113.

Meissner, W. (1998). Some notes on using walk-in traps. Wader Study Group Bulletin, 86, 33-35.

Meissner, W. (2007). To count or to catch? Do walk and in traps lead to a biased measure of wader productivity? Wader Study Group Bulletin, 113, 53–56.

Meissner, W. (2008). To count or to catch: a comparison of two methods of determining wader migration phenology. Wader Study Group Bulletin, 115, 16–19.

Meissner, W. (2009). A classification scheme for scoring subcutaneous fat depots of shorebirds. Journal of Field Ornithology, 80, 289-296.

Meissner, W., & Bzoma, S. (2011). Colour rings with individual numbers increase the number of ringing recoveries of small waders. Wader Study Group Bulletin, 118, 114–117.

Muraoka, Y., & Wichmann, G. (2007). Trap response of Wood Sandpipers, *Tringa glareola*. Ornis Fennica, 84, 140–142.

Pinchuk, P., & Karlionova, N. (2006). Use of playback calls for catching migrating Common Snipe *Gallinago gallinago* in autumn. Wader Study Group Bulletin, 110, 64–65.

Pollock, M.G., & Paxton, E.H. (2006). Floating mist nets: a technique for capturing birds in flooded habitat. Journal of Field Ornithology, 77, 335-338.

Prater, A.J., Marchant, J.H., & Vuorinen, J. (1977). Guide to the identification and ageing of Holarctic waders. Tring, UK: BTO.

Reynolds, C.M. (1986). A simple "head and bill" ruler. Wader Study Group Bulletin, 48, 12.

Selman, R.G., & Houston, D.C. (1996). A technique for measuring lean pectoral muscle mass in live small birds. Ibis, 138, 348–350.

Spencer R. (1972). The Ringer's Manual. Tring, UK: BTO.

Stanyard, D.J. (1979). Further notes on Curlew cramp and keeping cages. Wader Study Group Bulletin, 27, 19-21.

Stevens, M., Párraga, C.A., Cuthill, I.C., Partridge, J.C., & Troscianko, T.S. (2007). Using digital photography to study animal coloration. Biological Journal of the Linnean Society, 90, 211–237.

Svensson, L. (1992). Identification Guide to European Passerines. Stockholm, Sweden: L. Svensson.

Tree, A.J. (1972). Single shelf wader netting. Safring News, 1, 20-23.

Witherby, H.F., Jourdain, F.C.R., Ticehurst, N.F., & Tucker B.W. (1938-41). The Handbook of British Birds. (With corrections and additions 1943-44.) London, UK: Witherby.

List of Figures

List of Tables

Index

Bei Fragen zur Produktsicherheit wenden Sie sich bitte an:
If you have any questions regarding product safety,
please contact:

Walter de Gruyter GmbH
Genthiner Straße 13
10785 Berlin
productsafety@degruyterbrill.com